D1006447

The Greening of Industry

The Greening of Industry

A Risk Management Approach

EDITED BY
John D. Graham and
Jennifer Kassalow Hartwell

Distributed by
Harvard University Press
Cambridge, Massachusetts and
London, England
1997

Printed in the United States of America

10 9 8 7 6 5 4 3 2 1

This book is printed on acid-free paper, and its binding materials have been chosen for strength and durability.

Library of Congress Cataloging-in-Publication Data
The greening of industry ; a risk management approach / edited by John D. Graham and Jennifer Kassalow Hartwell.

p. cm.

Includes bibliographical references and index.

ISBN 0-674-36327-2 (cloth : alk. paper)

1. Health risk assessment — United States. 2. Environmental health — Government policy — United States. 3. Factory and trade waste — Health aspects — United States. I. Graham, John D. (John David), 1956– . II. Hartwell, Jennifer Kassalow, 1966–

RA566.27.G74 1997

615.9′02 — dc20 96-41873

CIP

CONTENTS

Preface

At the 1993 annual meeting of the Advisory Committee of the Harvard Center for Risk Analysis (HCRA), there was an intensive discussion about why risk analysis is often perceived by environmental organizations as a tool to legitimize decisions that would harm public health and environmental quality. Several committee members urged the HCRA to investigate the historical track record of risk analysis in the United States, to determine whether the poor perceptions of risk analysis were justified. This book represents the center's effort to be responsive to those concerns and suggestions.

Our working hypothesis, reflected in the project title (the "Green Industry" Project), was that risk analysis—as applied in federal regulation of industry throughout the 1980s and early 1990s—played a constructive role in promoting environmental health. We leave it to readers to judge whether our hypothesis is well supported by the evidence and whether our interpretations of recent history are insightful.

HCRA's Green Industry Project would not have succeeded without the persistent leadership of Jennifer Kassalow Hartwell, who served as staff director of the project from start to finish. She was not simply a coeditor of this book; she was the key person who made things happen.

Collaborative projects like this one are characterized by key events in which momentum is gained and big issues are resolved. A key event for us was a full-day retreat at the Endicott House in Dedham, Massachusetts, where chapter authors received comments from designated peer reviewers. The team of reviewers included Professor Richard Andrews (University of North Carolina School of Public Health), Professor Marc Landy (Boston University), and Professor John Evans (Harvard School of Public Health). The first and last chapters of the book, as well as the individual case studies, were improved dramatically as a result of this review process. We thank Professors Andrews, Landy, and Evans for their constructive suggestions.

During project deliberations, we found ourselves making frequent use of the risk management framework espoused by William Ruckelshaus in 1983. Consequently, we asked Mr. Ruckelshaus to read the first draft of the entire manuscript and provide comments. We appreciated his personal visit to the center in May 1996, at which point he made it clear that he felt we were indeed on the right track. We thank Mr. Ruckelshaus for taking the time to help us.

The authors of the individual chapters are credited for their roles. However, throughout the project each author was also asked to serve as an occasional peer reviewer of another author's chapter. There is no question that the quality of the final product was enhanced through this collaborative process.

Thanks are due to a variety of individuals who provided research assistance, contacts, and helpful data. Jennifer Kassalow Hartwell and John Graham appreciate the assistance of David Ailor, David Doniger, Phil Masciantoni, and Michael Shapiro in providing information for the coke chapter. Jim Hammitt and Kim Thompson thank Stephen DeCanio, A. McCulloch, Mack Farland, Pauline Midgley, Alan Miller, Mario Molina, Edward Parson, Tony Vogelsberg, Jonathan Weiner, and Donald Weubbles for helpful comments and discussions while preparing the ozone layer chapter. Milton Russell and Alvin Alm were extremely helpful with both information and comments for the chapter on lead in gasoline. Additional thanks go to Christopher DuMuth, Albert Nichols, Joseph Cannon, and William Ruckelshaus. Kim Thompson thanks Bill Seitz, Steve Risotto, Jon Meijer, Katy Wolf, George Smith, David Herzan, Ohad Jehassi, and John Evans for numerous helpful discussions and comments during several rounds of revision on the dry cleaning chapter, and Joe Bren, Jim Cogliano, Joe Cotruvo, William Johnson, Karen Levy, Archie McCullogh, Caffey Norman, Lorenz Rhomberg, Peter Voytek, and Amy Vasu for comments on the final draft. She also thanks the U.S. Environmental Protection Agency for partially supporting the development of the chapter through Cooperative Agreement CR818090-01-0, Program in Environmental Health and Public Policy. Kim Thompson and John Graham thank Bill Gillespie for numerous helpful discussions and Joan Abbott, Doug Barton, Peter DeFur, John Festa, Russell Frye, Thomas Jorling, Doug Pryke, and Mark Rossi for providing helpful comments and documents related to the pulp and paper chapter. We also thank

Danna Feldgoise, Gregory Goldfarb, and Susan Swedis for diligent research assistance as summer interns at the beginning of the project. We received helpful administrative assistance throughout the project from Mary Ester Otts, Ellen Patterson, Ken Repp, and Patricia Worden.

Special thanks are due to Dr. Andrew Smith for his technical assistance on several chapters as well as his "big picture" suggestions for how the project should proceed. Obviously this is a product that reflects the contributions of so many people that it is impossible to name everyone here. They know who they are, and let it be known that we are grateful to them.

John D. Graham, Ph.D.
August 1996

Contributors

Alison C. Cullen

Alison C. Cullen is an Assistant Professor at the Graduate School of Public Affairs at the University of Washington, where she teaches environmental policy processes and quantitative methods. Her research involves the analysis of environmental health risk, decision making in the face of risks that are uncertain or vary across populations, and the application of value of information and distributional techniques. Current projects include environmental exposure assessment in communities surrounding industrial facilities and sites in the United States as well as in Central and Eastern Europe. Professor Cullen was previously a faculty member at the Harvard School of Public Health and an environmental engineer in the Water Quality Branch of the U.S. Environmental Protection Agency. She has served as a technical consultant to the Natural Resources Defense Council and the Environmental Defense Fund, and she is on the Risk Assessment Advisory Committee for the state of California. She received her S.B. degree from MIT in civil and environmental engineering and M.S. and Sc.D. degrees in environmental health sciences from the Harvard School of Public Health.

Alan Eschenroeder

Alan Eschenroeder received B.M.E. and Ph.D. degrees at Cornell University in engineering. He is presently a Principal at Alanova, Inc., an environmental research and consulting firm in Lincoln, Massachusetts. Previously he was on the board of directors of the Gradient Corporation and was a senior consultant at Arthur D. Little, Inc., both in Cambridge, Massachusetts. His public service experiences range from serving as an elected city official in Santa Barbara, California, to chairing environmental study groups for the National Academy of Sciences. Professional assignments have focused on assisting communities in waste management decisions and analyzing environmental impacts of projects, using multimedia computer simulations. He has served as a technical advisor in Superfund mediation settlements and as an expert witness in court proceedings. Topics of his risk analyses include chemicals and human health, ballistic missile defense systems, industrial fog episodes, oil spills at supertanker terminals, and transportation of hazardous materials.

John D. Graham

John D. Graham is Professor of Policy and Decision Sciences at the Harvard School of Public Health, where he teaches the methods of risk analysis and cost-benefit analysis.

Professor Graham is the founding director of the Harvard Center for Risk analysis and the elected President of the Society for Risk Analysis, an international professional society of 2,500 scientists and engineers. Since joining the Harvard faculty in 1985, he has focused his research on toxic chemicals, automobile fuel economy, HIV infection, drinking and driving, and the methods of risk analysis. He is the author of six books, including *In Search of Safety: Chemicals and Cancer Risk,* and dozens of scientific articles. He serves on the international editorial board of the journals *Risk Analysis* and *Accident Analysis and Prevention.*

George M. Gray

George M. Gray is Deputy Director of the Harvard Center for Risk Analysis and Instructor in Risk Analysis at the Harvard School of Public Health. As a toxicologist, he is a strong proponent of using more and better scientific information in the risk-assessment process. His research interests include the interpretation of animal bioassay data for risk assessment, characterization of risk, and the use of risk information in corporate decision making on health, safety, and the environment. He has worked with many companies and trade groups, as well as federal and local government groups, to increase understanding of the size and sources of health and environmental risks. He holds a B.S. in biology from the University of Michigan and M.S. and Ph.D. degrees in toxicology from the University of Rochester.

James K. Hammitt

James Hammitt is Associate Professor of Policy and Decision Sciences at the Harvard School of Public Health, where he teaches decision theory and cost-benefit analysis. His research addresses the effects of risk and uncertainty on environmental policy choice. Current topics include evaluation of policies toward global-scale, long-term environmental threats like global climate change and stratospheric-ozone depletion; economic valuation of health risks and environmental quality, using revealed preference and contingent valuation methods; and assessment of the effect of financial risk due to Superfund liability on firms' cost of capital. Professor Hammitt was previously Senior Mathematician at RAND and a faculty member at the RAND Graduate School of Policy Studies. He received his A.B. and Sc.M. degrees in applied mathematics and his M.P.P. and Ph.D. in public policy from Harvard University.

Jennifer Kassalow Hartwell

Jennifer Kassalow Hartwell managed the Green Industry Project at the Harvard Center for Risk Analysis in 1994 and 1995. Prior to coming to Harvard, she was a research fellow at the Natural Resources Defense Council in New York where her work concentrated on ocean water quality and solid waste issues. Currently, Ms. Hartwell is pursuing a doctorate in organizational transformation in the Carroll School of Management at Boston College. Her research interests include the role organization behavior plays in industry's strategic environmental decisions. She earned her B.S. in environmental studies from the University of Vermont and her M.E.S. from the Yale School of Forestry and Environmental Studies.

Laury Saligman

Laury Saligman is currently a Commercial Development Analyst at Air Products and Chemicals, Incorporated, involved in new product research and development. She received her M.S. degree in environmental health sciences from the Harvard School of Public Health, and her undergraduate degree in physics from Georgetown University. She was a Research Fellow at the U.S. Environmental Protection Agency in the Office of Cooperative Environmental Management, where her area of focus included pollution prevention and the Toxic Release Inventory.

Kimberly M. Thompson

Kimberly Thompson holds B.S. and M.S. degrees in chemical engineering from the Massachusetts Institute of Technology, and a Doctor of Science degree in environmental science and engineering from the Harvard School of Public Health. She has applied a statistical decision-analytic approach to assess the value of better national exposure information for risk managers addressing the uncertain cancer risks posed by using perchloroethylene for dry cleaning. Her research interests include developing quantitative methods for risk assessment and risk management, and the implications of including uncertainty and variability in risk models.

The Greening of Industry

ONE

The Risk Management Approach

John D. Graham

Jennifer Kassalow Hartwell

Public policy toward environmental protection in the United States is now at a turning point. Governors, mayors, business leaders, and members of the scientific community have joined together in a powerful coalition, urging the U.S. Congress and the president to reform environmental laws and regulations. This unusual coalition shares a conviction that rational reform can achieve more protection of public health and the environment at less cost to the private and public sectors than is being accomplished by the patchwork of laws adopted since the first Earth Day was celebrated in 1970. The case for reform has been amply documented in reports by such diverse groups as the Clinton-Gore administration, the National Academy of Public Administration, the Carnegie Commission, the U.S. Office of Management and Budget and the General Accounting Office, the Business Roundtable, the National Research Council, and the Harvard Group on Risk Management Reform.

The precise directions of reform are not yet clear, but it appears that future environmental laws and regulations will be based on "risk management" — a series of principles and associated analytical determinations first promoted by former Environmental Protection Agency (EPA) administrator William Ruckelshaus (Ruckelshaus, 1983 and 1985). While risk management connotes different things to different people, there are some common themes in the various advocacy positions that favor risk management.

First, before regulating an alleged environmental problem, an objective assessment of scientific knowledge must indicate that exposure to the pollutants of concern may represent a significant danger to human health or the environment (*IUD v. API,* 1980). Although the danger need not be proven or known with certainty, the best available scientific information must provide a plausible basis for concern. This principle creates a scientific hurdle, but not one that is so high that it blocks reasonable precautionary regulations.

Second, to make efficient use of resources, environmental problems should be ranked in order of priority by some formal or informal "comparative risk" process. Other things being equal, the more serious problems should be given greater priority than the less serious ones (EPA, 1987; EPA, 1990; Davies, 1996; Finkel and Golding, 1994).

Third, the proposed regulations should reduce the risks of targeted pollutants by a greater degree than they increase other risks (pollution-related and otherwise) to public health and the environment. This consideration of "risk trade-offs" requires some broad-based assessment of the consequences — both unintended and intended — of regulatory action (Warren and Marchant, 1993; Graham and Wiener, 1995).

Finally, the economic costs of contemplated actions must be reasonably related to the degree of risk reduction expected from the reduction in pollution — an explicit balancing of costs and benefits (Freeman, 1993; Tolley, Kenkel, and Fabian, 1994). The cost-benefit analysis may be quantitative and monetary (where possible), but in many cases it may be more qualitative and judgmental (Portney, 1988).

In the political marketplace, these principles are often advocated with the following shorthand: "good science," "risk-based priorities," "risk trade-offs," and "cost-benefit." To some observers these four principles may appear to be nothing more than a restatement of common sense. It is important to realize, however, that these principles represent a major departure from the philosophy found in many current environmental laws. The reform principles are based more on utilitarian reasoning than on notions of "rights" to environmental protection or egalitarianism (Schroeder, 1986).

For decades Congress has passed environmental laws on the assumptions that pollution is morally wrong, that citizens are "victims," and that therefore pollution should be eliminated — or at least reduced until public health and the environment are protected — regardless of the cost (Dwyer, 1990). Some current laws permit regulators to make practical judgments, but even these laws may require pollution to be reduced to the lowest feasible levels regardless of the magnitude of risk reduction or the incremental costs of achieving reductions in pollution (*ATMI v. Donovan,* 1981; McGarity, 1991). The patchwork of existing laws provides for none of the broad-based risk analysis that excites reform advocates (Lave and Males, 1989).

Despite a broad bipartisan consensus that some policy reforms based on risk management are needed, the official position of most national and grassroots environmental advocacy groups is that such reforms will "weaken" environmental protection. They vigorously oppose such reforms out of fear that promising pollution prevention policies cannot be defended on the basis of risk management. Some advocates have gone so far as to suggest that risk management will cause a "rollback" in the environmental progress that has been achieved under current laws. And some groups advocating risk-based reforms may be doing so in the hope that such reforms will make it difficult or impossible for the federal government to regulate industrial pollution.

This book examines the question of whether reforms based on risk management can reduce industrial pollution. Our approach is empirical, rooted in an assessment of historical experience where the federal government chose to employ risk management, even though existing laws may or may not have encouraged such thinking. The basic argument of the book is that risk management can support pro-environment policies at the same time that it discourages or blocks poor policies (that is, those that would impose substantial costs with few returns in environmental protection). Those who expect risk management to eliminate or radically curtail environmental regulation may be in for a big surprise. Far from undermining environmental policy, risk management may serve to strengthen the credibility of industrial regulation by giving it a stronger basis in science and economics, with less dependence on moral or legalistic claims. If this perception is correct, it suggests that professional environmentalists and their allies among elected officials should reconsider their opposition to the concept and focus on how risk management can be deployed to advance environmental protection.

By suggesting that risk management has promise, we do not deny the remarkable environmental progress that has been achieved in the past thirty years under existing laws (Portney, 1990; CEQ, 1990). Concentrations of various pollutants in the air are declining. The fraction of lakes and rivers that are fishable and swimmable is on the rise. While the volume of hazardous wastes produced in the country continues to rise, more care is being devoted to how these wastes are transported, stored, and managed. And there is a growing public awareness, reflected in at least some trends in the day-to-day behavior of households and businesses, that more effort is being devoted to preventing pollution at its source, rather than waiting to clean it up after the problem has been created.

Despite these encouraging trends, new approaches will be necessary to solve the nation's remaining environmental problems (NAPA, 1995). Our nation's residual pollution problems tend to be less obvious and more complex than was the case in 1970. The annual cost of compliance with environmental regulations is increasing rapidly in both the public and private sectors. Serious concerns have been expressed about whether resources are being directed at the most serious remaining problems. For example, the speculative dangers posed by asbestos in school buildings and toxic chemicals in the soil at abandoned waste disposal sites have captured media and government attention, even though more serious potential problems, such as indoor radon and global climate change, have received less public attention, or are less readily comprehended (Breyer, 1993; Cross, 1990 and 1994).

Belief in risk management emerged from the practical experiences of a succession of dedicated public servants who struggled to lead the U.S. Environmental Protection Agency under the Carter, Nixon, Ford, Reagan, and Bush administrations. It is useful, therefore, to place the emergence of risk management in some historical context.

| William Ruckelshaus: Ahead of His Time?

Since Earth Day of 1970, the American people and their elected representatives have supported a strong role for the federal government in protecting people and the natural environment from pollution generated by industrial activity. The predominant public view in the 1970s was that more government regulation was needed to protect people and the environment from pollution by business (Dunlap, 1991).

While liberal Democrats tend to be strong advocates of environmental regulation, the EPA was created through executive action by President Nixon in 1970. William Ruckelshaus (attorney, Indiana legislator, unsuccessful Republican candidate for the U.S. Senate, and Justice Department official) was appointed the agency's first administrator. Ruckelshaus shared the prevailing pro-environmental viewpoint, which led to some highly controversial decisions, such as his 1972 ban of the pesticide DDT. To this day, Ruckelshaus praises the law under which DDT was banned, the Federal Insecticide, Fungicide, and Rodenticide Act, because it authorized the EPA to engage in risk management by weighing the risks of the pesticide against its benefits and those of substitutes. When Ruckelshaus left the EPA in 1973, his successor, Russell Train, continued to make tough regulatory decisions aimed at protecting the environment.

Passing new environmental laws was a popular course for politicians in the 1970s. With huge bipartisan majorities, Congress passed numerous such laws, including the Clean Air Act (1970), the Clean Water Act (1972), the Safe Drinking Water Act (1974), the Resource Conservation and Recovery Act (1976), the Toxic Substances Control Act (1976), and the Superfund law governing cleanup of abandoned hazardous waste sites (1980). As it pursued its mandate to make life safer and healthier for humans and nonhuman species, the EPA drew on the legal powers in these laws to control some of the day-to-day behaviors of private citizens as well as small and large businesses.

When President Carter appointed Douglas Costle, a former state regulator, to succeed Train at the EPA in early 1977, the stage seemed to be set for an even more aggressive regulatory period. It was the Carter administration, however, that was the first administration to begin to ask serious questions about the future of environmental policy. Faced with the Iranian oil crisis, President Carter's search for a national energy and economic policy ran into conflict with his commitments to environmental protection. How could domestic production of coal and oil be increased rapidly if these industries were restrained by environmental regulations? Ultimately, the Carter administration resorted to White House oversight of EPA activities to mediate conflicting policy goals, particularly as the national economy entered a severe recession in 1980. Inside the EPA, early work on risk-analysis methods was being done at the Carcinogen Assessment

Group, the Office of Policy Analysis, and some program offices. But more important, Ronald Reagan's victory over President Carter in the 1980 presidential election led to a drastic shift in federal environmental policy.

Under President Reagan's leadership, substantial cuts were made in the EPA's budget, while power to oversee the agency's regulatory ambitions was consolidated in the Office of Management and Budget. Building on previous executive orders, Reagan's Presidential Executive Order 12291 required that major new regulations be supported by cost-benefit analysis. But instead of taking an analytical approach, Reagan's initial appointees to the EPA were generally hostile to implementation of strict environmental laws (regardless of benefits and costs). This antiregulatory position was not sustainable. The Reagan administration was repeatedly accused of being more interested in delivering "regulatory relief" for industry — a promise that had been prominently voiced by the 1980 Reagan campaign — than in developing a sound approach to environmental protection. This lack of a principled policy framework, coupled with charges of corruption and incompetent management at the EPA, soon created a major political problem for the Reagan administration (Landy, Roberts, and Thomas, 1990).

To rescue his administration's environmental policy, Reagan turned to William Ruckelshaus, the EPA's first and former administrator from 1970 to 1973, who had been working in private industry since his departure from the agency. President Reagan asked Ruckelshaus to return to the helm of the EPA and restore public confidence in the federal government's environmental policies.

Facing the dilemma of enormous legal responsibilities, public mistrust of the EPA, and a fragile national economy, Ruckelshaus proposed risk management as a principled framework for environmental policy (Ruckelshaus, 1983 and 1985; EPA, 1984). Ruckelshaus saw science as the tool, the means to identify the significant environmental threats (risk assessment), while the methods of policy analysis (such as cost-benefit analysis) would be used to determine when the risks of pollution were reduced to acceptable levels (risk management). He saw the laws governing pesticide regulation, which called for discretionary regulation based on "unreasonable risk" determinations, as a model for the reform of other environmental laws.

The tool of risk analysis was certainly not invented by Ruckelshaus. It had already been used for decades in the nuclear power and food industries (HCRA, 1994). The contribution made by Ruckelshaus was the coherent statement of how risk management could bring professionalism and rigor to environmental policy. Ruckelshaus was notably unsuccessful, however, in persuading a Democratic Congress to rewrite environmental laws based on risk management. Looking back, it appears that his vision of environmental policy was more than ten years ahead of what is now transpiring. Nevertheless, Ruckelshaus did launch new scientific and analytical efforts in programs throughout the agency, efforts that are the subject of several case studies in this book.

Ruckelshaus's successors at the EPA, Lee Thomas and William Reilly, reinforced the EPA's commitment to risk management. Thomas, a career public servant, launched a program of comparative risk assessment, as senior scientists and managers in the agency ranked the remaining environmental problems facing the United States (EPA, 1987). They found that the problems of most concern to scientists (for example, global warming) were different from the problems of most concern to Congress and the public (such as hazardous waste sites).

Reilly, a well-known conservationist, refined this risk-ranking process in collaboration with EPA's Science Advisory Board, a group of independent scientists from universities and the private sector (EPA, 1990a). He then marketed comparative risk analysis as a priority-setting tool to Congress, governors, and mayors. Reilly has commented that during his four years at the helm of the EPA (1989–1992), the fraction of the agency's budget devoted to "high-risk" dangers increased from 15 percent to 30 percent (Reilly, 1994).

The direct cost to taxpayers of operating the EPA is relatively small (about $7 billion in fiscal year 1995) compared with the trillion-dollar budgets of the Department of Defense and the Department of Health and Human Services. However, EPA programs impose a large cost on the operation of the private sector and on states and localities. The EPA estimates that the costs of complying with federal environmental laws were about $115 billion in 1990 and will exceed $150 billion by the year 2000 (EPA, 1990b). Senator Daniel Patrick Moynihan, the senior Democrat from New York, has observed that these expenditures are not necessarily too much to spend for environmental protection but they are certainly too large to spend unwisely (Moynihan, 1993).

By the time President Clinton entered the White House, the intellectual case for risk management was gaining momentum. Key Democrats, notably Senator Moynihan of New York and Bennett Johnston of Louisiana, began to introduce bills in Congress to promote a risk-based approach to environmental policy. When the Clinton-Gore administration tried to elevate the EPA to cabinet status in 1993, they were surprised to learn that this could be passed in the House and Senate only if the administration were willing to accept an amendment promoting the role of risk analysis in the new cabinet-level agency. At the urging of organized environmentalists, the administration chose not to press the issue of cabinet status, and the proposal died.

The Republican takeover of Congress in 1994 accelerated political interest in reform of environmental policy as the "Contract with America" written by House Republicans included an extensive endorsement of the principles of risk management. The Senate, however, did not pass broad-based regulatory reform legislation based on the principles of risk management, and instead has begun to incorporate the principles of risk management into each major environmental law as it is considered for reauthorization (e.g., the Safe Drinking Water Act). The Clinton administration has opposed the broad-based reforms as a "rollback" of environmental protection, because of their excessive procedural and legalistic requirements (Dowd, 1994). However, the administration has acknowledged the merits of risk management as a tool in environmental policy and has its own initiatives in this area (Katzen, 1995).

| Opposition to Risk Management

Swimming against the tide are national and grassroots environmental advocacy groups, which are generally united in their skepticism about a greater role for risk management in environmental protection (Silbergeld, 1993). From their perspective, the past thirty years of environmental policy have been fairly successful, as major progress has been made in cleaning up the air, water, and the food supply. While much more progress could have been made, they fear that risk-based reforms of environmental laws will ultimately result in less rather than more environmental protection. Instead they often advocate more substantial reforms that are aimed at eliminating the structural causes of pollution in the industrial economy (Hornstein, 1993; O'Brien, 1994).

Many environmentalists accept the usefulness of qualitative aspects of risk assessment (that is, using science to identify those pollutants that are potentially hazardous to people or wildlife). They advocate, for example, bans or reductions in the use of chemicals or products that are judged by scientists to be potentially toxic or hazardous. Some environmentalists even accept quantitative risk assessment as a regulatory tool, but most see little value in speculative efforts to compare risks, or to express the benefits and costs of regulation in dollar units (Latin, 1988).

The stated reasons for opposition to quantitative risk assessment vary. Some see this analytical approach as elitist and technocratic, since it is difficult for nonexperts to participate in policy discussions framed by the analytical tools of risk assessment (Commoner, 1989 and 1995). Others see the approach as ethically or morally suspect, since it ultimately permits some industrial pollution to persist in situations where the remaining risks of pollution are estimated to be small or the costs of further pollution control are judged to be too high (Doniger, 1989). Insofar as any pollution of the natural environment is morally suspect, then any risk-based determinations that permit pollution are morally suspect. And some fear that science has not progressed enough to inform a risk-based approach, leading to the conclusion that policy should be based on the "better safe than sorry" principle. For example, risk assessors tend to be better at quantifying human health effects than ecosystem effects and often focus exclusively on cancer risks, virtually ignoring possible risks to the immune, developmental, reproductive, and neurological systems in the human body (Silbergeld, 1993).

While each of these objections is important, the most practical objection to risk-based reforms is that they will not be as effective in cleaning up the environment as implementation of current or stricter laws would be (Prendergrass, Locke, and McElfish, 1995). Current environmental laws often authorize ambitious regulatory programs based on only qualitative determinations of potential hazard. These laws force industry to adopt specific pollution control technologies or cleaner processes with limited or no consideration of the amount of risk reduction involved or the costs of these changes. Change is mandated by law and is not subjected to discretionary analysis by regulators.

Risk-based reforms are seen as a device to raise the burden of proof for regulatory action. Before acting, a regulator might have to go beyond a qualitative determination of hazard and show that 1) a targeted risk is significant based on quantitative analysis, 2) the risk is a priority compared with other risks that have not yet been addressed, 3) the contemplated action will not create unacceptable new risks, and 4) the costs of risk reduction are not excessive compared with the difficult-to-measure benefits of risk reduction.

By raising these analytical hurdles, risk management is seen as causing a slower regulatory response to environmental problems (Browner, 1995). Accumulation of data to support risk assessments can proceed indefinitely. New uncertainties continually arise, and decisions tend to be postponed until better data are available. The demand for risk analysis can be used as a delay tactic by industry and regulators, thereby allowing polluters to escape regulatory control (Citizen Action Guide, 1994).

Comparing risks is seen as a defeatist position, since it is based on the premise that efforts cannot be made to reduce all risks of pollution (O'Brien, 1994). The lack of enthusiasm among environmentalists for risk-based priority setting should not be surprising. They choose to advocate increases in the resources available for environmental protection, rather than participating in the rationing of scarce resources or triaging important problems (Hornstein, 1992).

Cost-benefit analysis, while seen as a tool to assist in running a business, repels environmentalists as a tool of environmental policy (Greer, 1995). They believe that the costs of environmental protection are easier to quantify and express in dollars than are the benefits of environmental protection. After all, what are the economic values of feeling healthier, living longer, being able to see farther and more clearly on a summer day, knowing that clean groundwater will be available for use by future generations, and protecting endangered species from extinction? There is also a suspicion that the costs of preventing or controlling pollution are not as large as industry and regulators suggest, and there are certainly cases in which industry has exaggerated (consciously or inadvertently) the costs of complying with proposed regulations. Unless and until these questions and concerns can be answered persuasively, environmentalists will certainly oppose a greater role for cost-benefit analysis in environmental policy.

In 1994 the leaders of the nation's major environmental groups joined together to urge Congress and the public to resist the "anti-environmental arguments that industry is using." They highlighted industry's "misleading" advocacy of comparative risk analysis and cost-benefit analysis. They referred to these analytical tools as "lies" being used to weaken environmental laws (Citizen Action Guide, 1994). President Clinton's EPA administrator, Carol Browner, has used similar rhetoric, suggesting that risk-based policy would never have permitted the phase-out of lead in gasoline. The price, she warns, would have been millions of permanently brain-damaged children (Browner, 1995).

| What Is the Track Record of Risk Management?

Are risk management and environmental protection compatible? Forecasts about the practical consequences of risk-based reform of environmental policy need not be made on a purely theoretical or conceptual level. The United States now has had more than a decade of experience since Ruckelshaus returned to the helm of the EPA in 1983 and began to promote risk management through administrative reforms. Although many environmental laws do not require or encourage the use of risk management, the EPA has still made extensive use of this intellectual approach since the early 1980s. This book aims to assess this experience in order to better understand the potential impact of more fundamental risk-based reforms of environmental legislation.

A number of important questions need to be asked about the EPA's experience with risk management. How and why are risk assessments initiated? Who conducts them, and what are the sources of information about risk? Have industry leaders and environmentalists accepted the findings of these risk assessments, and if not, have they persuaded the EPA and the scientific community that the EPA's position is untenable? Do cost-benefit analyses support regulation, and if so, how are the economic benefits of environmental regulation quantified? Are cost-benefit studies an exercise in burdensome paperwork, or do they play an important role in persuading actors that regulation is or is not necessary? What specific steps have the EPA and targeted industries taken to reduce or prevent pollution? Where progress has been made, are factors other than risk management partly or wholly responsible for the progress? The aim of this volume is to provide some empirical evidence relevant to answering these questions.

| Methods

To better understand the role of risk management, we have elected to perform in-depth case studies of particular industries. The six industrial processes or products we selected for study are leaded gasoline, perchloroethylene use in dry cleaning, coke-based steelmaking, incineration of municipal waste, chlorine-based pulp and paper production, and chlorofluorocarbon use. They were selected in part because our working group at the Harvard Center for Risk Analysis was knowledgeable about them, but also because they were major targets of EPA risk assessments in the 1980s. In each case, sufficient time has elapsed since the publication of important risk assessments to determine at least the initial responses of regulators and industry.

The case studies explain the nature of the pollution problems occurring in each industry and the scientific basis for concerns that such pollution is harmful to human health or the natural environment. We also provide the reader with basic information about the prevailing technologies used in the industry as well as the pace of innovation. Where necessary, the industry is broken down by segments or firms to clarify important differences within the industry.

Each case study is based on publicly available information and personal or telephone interviews with key individuals in the public and private sectors. Information is presented in a descriptive fashion to allow readers to form their own judgments about cause and effect, and what the normative implications of a case might be for reform of environmental policy. While we have attempted to present the case information in as neutral a fashion as possible, we recognize that no presentation of information on such complex and controversial matters can be completely unbiased. We have, however, reserved our evaluative comments for the conclusion of each case chapter and for the last chapter, where we attempt to raise some general hypotheses and draw some crosscutting insights.

For a variety of reasons, the cases should not be considered as strictly parallel examinations of the same questions. While some of the cases focus on the fate of particular products (such as leaded gasoline), other cases focus on particular methods of industrial production (chlorine-based pulp and paper production; coke-based steelmaking). Moreover, some of the cases find the EPA working to resolve a potentially global environmental problem (for example, chlorofluorocarbon use as a threat to the stratospheric ozone layer), while other cases reveal EPA involvement in issues with a strongly local character (the use of municipal incinerators). A particular method of analysis (such as cost-benefit analysis) plays a prominent role in some cases but not in others.

Nor is the process of analysis in the case studies homogeneous. A risk assessment may cover only the projected dangers that might occur in the absence of regulation, while a regulatory impact analysis might predict the changes in levels of risk and the economic costs resulting from alternative regulatory options (including no action). Some analyses are initiated by government (for example, those done to support new national regulation of the paper industry), while others are initiated by industry (such as those undertaken to obtain a permit for a hazardous waste incinerator). Some analyses address only human health considerations, while others include ecological impacts as well. Sources of information for analyses also vary: sometimes government collects original toxicity or exposure data for use in a risk assessment; in other cases these data may be supplied by corporations or consultants hired by trade associations.

The case study method of research is interesting and useful but not always as definitive as some other forms of scientific analysis. Readers should not expect unequivocal conclusions from the cases in the way a toxicologist looks for a dose-response relationship in an experiment or an economist determines the statistical relationship between the rate of inflation and the size of the federal deficit. The case method is best used to learn about the institutional context for decisions and to develop hypotheses about what might or might not have been occurring in that context (Graham, Green, and Roberts, 1988). By itself, the case study cannot prove the "counterfactual" (it cannot show what would have happened to the paper industry if the EPA's dioxin risk assessments at paper mills had not been conducted). Provided with the contextual information from the case studies, we urge readers to formulate their own hypotheses as we offer some that occur to us.

A strength of this book is the rich diversity of industries and products that are addressed. Our selection of cases was certainly not random, and thus we make no claim that these six cases are representative of some unspecified universe of industries and products that might be targets of EPA regulation. Indeed, we have not conducted any case studies of industries where the EPA has never performed a risk assessment. We believe there is merit in judging Ruckelshaus and his successors on what happened to pollution when risk management was employed, since the future policy question will be whether the framework should be applied more broadly or even universally.

| Sources

American Textile Manufacturers Institute, Inc. v. Donovan, (ATML v. Donovan), 1981. 452 U.S. 5490.

Breyer, S., 1993. *Breaking the Vicious Circle: Toward Effective Risk Regulation* (Cambridge: Harvard University Press).

Browner, C., 1995. Letter from Carol Browner, EPA Administrator, to Hon. George Brown, House Committee on Science, Washington, D.C., January 31, 1995.

Carnegie Commission on Science, Technology, and Government, 1993. *Risk and the Environment: Improving Regulatory Decisionmaking* (Washington, D.C.).

Center for Risk Analysis, 1994. *A Historical Perspective of Risk Assessment in the Federal Government* (Boston: Harvard School of Public Health), pp. 29–31, 38–48.

Citizen Action Guide, 1994. *How to Defend Our Environmental Laws,* compiled by American Oceans Campaign and twelve other groups (Washington, D.C.).

Commoner, B., 1989. "The Hazards of Risk Assessment," *Columbia Journal of Environmental Law* 14: 365–378.

———, 1994. "Pollution Prevention: Putting Comparative Risk Assessment in Its Place," in A.M. Finkel and D. Golding (eds.), *Worst Things First? The Debate over Risk-Based National Environmental Priorities* (Baltimore: Johns Hopkins University Press), pp. 203–228.

Council on Environmental Quality, (CEQ), 1990. *Environmental Quality: 21st Annual Report* (Washington, D.C.: Government Printing Office, table 39.

Cross, F. B., 1990. *Legal Responses to Indoor Air Pollution* (New York: Quorom Books).

———, 1994. "The Public Role in Risk Control," *Environmental Law* 24:888–969.

Davies, J. C., 1996. *Comparing Environmental Risks: Tools for Setting Government Priorities,* (Washington, D.C.: Resources for the Future).

Doniger, D., 1989. National Clean Air Coalition, "Clean Air Act Amendments of 1989," *Hearings before the Senate Committee on Environment and Public Works,* 101st Cong., 1st sess., September 21, 28–30.

Dowd, A. R., 1994. "Environmentalists Are on the Run," *Fortune,* September 19, pp. 91–100.

Dunlap, R. E., 1991. "Trends in Public Opinion toward Environmental Issues: 1965–1990," *Society and Natural Resources* 4:285–312.

Dwyer, J. P., 1990. "The Pathology of Symbolic Legislation," *Ecology Law Quarterly* 17:233–316.

Environmental Protection Agency (EPA), 1984. *Risk Assessment and Management: Framework for Decision Making* (Washington, D.C.: Environmental Protection Agency).

———, 1987. *Unfinished Business: A Comparative Assessment of Environmental Protection* (Washington, D.C.: Environmental Protection Agency).

———, 1990a. *Reducing Risk: Setting Priorities and Strategies for Environmental Protection* (Washington, D.C.: Environmental Protection Agency).

———, 1990b. *Environmental Investments: The Cost of a Clean Environment* (Washington, D.C.: Environmental Protection Agency).

Finkel, A. M., and D. Golding (eds.), 1994. *Worst Things First? The Debate over Risk-Based National Environmental Priorities* (Washington, D.C.: Resources for the Future).

Freeman, A. M., 1993. *The Measurement of Environmental and Resource Values: Theory and Methods* (Washington, D.C.: Resources for the Future).

Graham, J. D., L. Green, and M. J. Roberts, 1988. *In Search of Safety: Chemicals and Cancer Risk* (Cambridge: Harvard University Press).

Graham, J. D., and J. B. Wiener (eds.), 1995. *Risk versus Risk: Tradeoffs in Protecting Health and the Environment* (Cambridge: Harvard University Press).

Greer, L. E., 1995. Senior Scientist, NRDC, "The Role of Risk Assessment and Cost-Benefit Analysis in Regulatory Reform," *Hearings before Senate Committee on Government Operations,* 104th Cong., February 15, 1995.

Harvard Center for Risk Analysis, (HCRA), 1994. *A Historical Perspective on Risk Assessment in the Federal Government* (Boston: Harvard School of Public Health).

Harvard Group on Risk Management Reform, 1995. "Special Report: Reform of Risk Regulation: Achieving More Protection at Less Cost," *Human and Ecological Risk Assessment* 1(3):183–206.

Hornstein D. T., 1992. "Reclaiming Environmental Law: A Normative Critique of Comparative Risk Analysis," *Columbia Law Review* 92:562–633.

———, 1993. "Lessons from Federal Pesticide Regulation on the Paradigms and Politics of Environmental Law Reform," *Yale Journal on Regulation* 10:369–446.

Industrial Union Department, AFL-CIO v. American Petroleum Institute, (IUD v. API), 1980. 448 U.S. 607.

Katzen, S., 1995. "Risk Assessment and Cost-Benefit Analysis for New Regulations," *Hearings before the House Committee on Commerce,* 104th Cong., 1st sess., February 1–2, 202–207.

Landy, M. K., M. J. Roberts, and S. R. Thomas, 1990. *EPA: Asking the Wrong Questions* (New York: Oxford University Press).

Latin, H., 1988. "Good Science, Bad Regulation, and Toxic Risk Assessment," *Yale Journal on Regulation* 5:89–148.

Lave, L. B., and E. H. Males, 1989. "At Risk: The Framework for Regulating Toxic Substances," *Environmental Science and Technology* 23:386–391.

McGarity, T. O., 1991. *Reinventing Rationality: The Role of Regulatory Analysis in the Federal Bureaucracy* (Cambridge: Cambridge University Press).

Moynihan, D. P., 1993. "Environmental Risk Reduction Act," *Congressional Record,* January 21, S550.

National Academy of Public Administration, (NAPA), 1995. *Setting Priorities, Getting Results,* Summary Report, April (Washington, D.C.: NAPA).

O'Brien, M., 1994. "A Proposal to Address, Rather than Rank, Environmental Problems," in A. M. Finkel and D. Golding (eds.), *Worst Things First? The Debate over Risk-Based National Environmental Priorities* (Washington, D.C.: Resources for the Future), pp. 87–105.

Prendergrass J., P. Locke, and J. McElfish, 1995. "The Environment and the Contract," *Environmental Law Review* 25:10350–10366.

Portney, P. R., 1988. "Reforming Environmental Regulation: Three Modest Proposals," *Issues in Science and Technology* 4:74–81.

——— (ed.), 1990. *Public Policies for Environmental Protection* (Washington, D.C.: Resources for the Future).

Reilly, W., 1994. Address to the John F. Kennedy School of Government, Harvard University, December 13.

Ruckelshaus, W., 1985. "Risk, Science, and Democracy," *Issues in Science and Technology* 1:10371.

———, 1983. "Science, Risk, and Public Policy," *Science* 221:1026–1028.

Schroeder, C. H., 1986. "Rights against Risks," *Columbia Law Review* 86:495–562.

Silbergeld, E. K., 1993. "Risk Assessment: The Perspective and Experience of U.S. Environmentalists," *Environmental Health Perspectives* 101 (June): 100–104.

Tolley, G., D. Kenkel, and R. Fabian, 1994. *Valuing Health for Policy: An Economic Approach* (Chicago: University of Chicago Press).

Warren E. W., and G. E. Marchant, 1993. "More Good than Harm: A Hippocratic Oath for Environmental Agencies and Courts," *Ecology Law Quarterly* 20:379–440.

TWO

The Demise of Lead in Gasoline

George M. Gray
Laury Saligman
John D. Graham

It was a case in which the benefits clearly outweighed the costs. The proposed regulation had sailed through the rulemaking process and review at the Office of Management and Budget (OMB) in just seven months. The rule was certain to improve public health. It provided the "deregulatory" Reagan administration an opportunity to demonstrate its commitment to protecting human health and the environment. It also represented a victory for the Environmental Protection Agency (EPA), which had suffered a number of demoralizing fiascoes. According to EPA officials, the idea had come from inside the EPA with little or no pressure from environmental or industry groups (Alm, 1994; Russell, 1994). Rather than being another example of "regulatory relief" for industry, as first envisioned, the plan was to actually increase the stringency of a regulation. So, in March 1985 William Ruckelshaus, administrator of the U.S. Environmental Protection Agency, decided that, rather than allowing the amount of lead used in gasoline to decrease slowly as the number of cars using leaded gasoline gradually dropped to zero, the EPA would require a more than 90 percent reduction in the amount of lead allowed in leaded gasoline. In addition, the use of lead in gasoline would be completely eliminated by 1995.

This chapter describes a case in which quantitative analysis, including assessment of risks to human health and formal cost-benefit analysis, played a central role in protecting and strengthening a significant piece of environmental regulation. The accelerated phaseout of lead in gasoline was a far-reaching regulation, affecting virtually all Americans, yet it was put in place with remarkable speed and little media attention. And, as we shall see, it was quantitative policy analysis, performed with no statutory deadline and no internal or external pressure, that provided the impetus for the regulation.

| Why Put Lead in Gasoline?

Lead was first discovered to be an octane booster for fuels by Thomas Midgely and his colleagues at General Motors in 1922, as automobile and oil companies tested thousands of potential octane boosters and antiknock agents (Rosner and Markowitz, 1985). It was really the only viable material identified. By increasing gasoline's octane rating, the presence of lead improves engine power and fuel economy while at the same time avoiding the problem of preignition (knocking). Lead was considered the ideal additive because only a small amount was needed. For example, the addition of 0.5 grams of lead per liter of gasoline increases octane by about 5 points on a 100-point scale. The lead used in gasoline is chemically bound to carbon molecules and, therefore, is an organic lead compound. The compound was patented, and until 1948 the Ethyl Corporation was the exclusive supplier of tetraethyl lead to the oil refining industry.

An early incident that raised concerns about the health effects of lead in gasoline occurred in October 1924. In the experimental laboratories of the Standard Oil Company, out of forty-nine workers, five workers died and thirty-five experienced severe neurological symptoms from organic lead poisoning. This event raised concerns about exposure of the general population to organic lead compounds in automobile exhaust. In fact New York state and Philadelphia and other municipalities briefly banned the sale of leaded gasoline after this event. After the furor died down, the use of lead in gasoline returned and continued until the early 1970s.

Over the years, many other companies tried to identify alternatives to compete with tetraethyl lead, but none panned out. Oil producers and refiners increased the octane rating of the gasoline with changes in the refining process, but improvements in car engine technology over time meant that the octane boost from lead was still necessary. By the early 1970s more than 200,000 tons per year of lead were being added to gasoline.

| Lead and Health

All of the lead that is taken into the body is not absorbed. Depending on the form of the lead, up to 15 percent of an ingested dose in adults is absorbed in the intestines. Organic forms of lead, such as tetraethyl lead, are absorbed more readily than inorganic forms. Children may absorb as much as 40 percent of an oral dose of lead. The rate of lead absorption by the lungs may be even higher. Certain conditions, including pregnancy or a lack of calcium in the diet, can increase the body's absorption of lead. Once in the body, about 95 percent of lead is stored in hard tissues such as bones and teeth, with the rest distributed among the soft tissues and blood. Scientists have come to rely on the blood lead level as an accurate, easily measured indicator of the body's lead burden. The blood lead level is expressed in micrograms per deciliter (μg/dl). One microgram is about 3 ten-millionths of an ounce of lead; a deciliter is about 3.4 ounces of blood.

Adverse effects of lead on human health have been suspected since ancient times, and the toxicological profile of lead has been built up over the years (Goyer, 1991). At the beginning of the twentieth century, reports from around the world began to point to lead as a significant neurotoxicant. The most sensitive effects of lead, those that occur at the lowest doses, are centered on the nervous system. In adults, lead may have effects on the brain, with changes in behavior or increased fatigue at relatively low exposure levels. Higher levels of exposure are associated with damage to the peripheral nervous system, responsible for sensation and motor control. In fact painters' wrist-drop, due to damage to the long neurons that control the muscles of the hand and forearm, was a well-known occupational hazard for house painters who used lead-containing paint.

At even higher doses, lead is associated with a number of other toxic effects, including anemia, due to inhibition of enzymes involved in hemoglobin synthesis; effects on the endocrine system; kidney toxicity; and developmental damage. Reports of behavioral and learning deficits in children exposed to relatively low levels of lead first appeared in the 1940s. The concern was that young children might be more sensitive to the neurotoxic effects of lead because their "blood-brain barrier" is incompletely developed and lead can more easily reach the brain.

The government began to be concerned about the effects of lead on children in the late 1960s. At that time, between 0 and 60 μg/dl was considered an acceptable blood lead level for children. In 1970 the U.S. surgeon general recommended that screening be undertaken to identify children with blood lead levels higher than 40 μg/dl. Screening showed that, in some cities, 20 to 45 percent of all children had blood lead levels in excess of 40 μg/dl (Needleman, 1992).

In the early 1970s scientists believed that the primary sources of childhood lead exposure were deteriorating lead paint in homes, lead in food (from lead solder in cans), lead in water (from lead pipes and joints), and environmental sources such as air and dust. There was some disagreement about the relative contribution of each source to children's blood lead levels, but studies were beginning to point to environmental sources as significant contributors to blood lead levels.

The evolving state of knowledge about the potential health effects of lead played a large role in the regulation of leaded gas. It was generally agreed in the early 1970s that there was some evidence that lead could produce neurological effects in children at levels below those causing clinical symptoms of poisoning. At that time, however, the evidence was still considered preliminary and the size of the effect highly uncertain. Through the 1970s and early 1980s an increasing number of studies examined the effects on children of low levels of lead exposure. These epidemiological studies used more sophisticated designs and were conducted with much larger study populations (Smith, Grant, and Sons, 1989). The crucial finding was that childhood lead exposure was associated with decrements in cognitive function, usually measured as IQ (Rummo et al., 1979; Needleman et al., 1979; Yule et al., 1981).

Three important issues were raised by the many studies conducted. First, adverse effects on cognitive development and function were seen at blood lead levels far below those necessary to generate clinical symptoms. Small IQ decrements were observed in children with blood lead levels in the neighborhood of 20 μg/dl, perhaps even lower. Second, studies were not demonstrating a clear threshold for the effects of lead on children's IQ scores. That is, it was impossible to estimate a level of lead exposure, or blood lead level, that was without risk of cognitive impairment. Third, the effects of lead appeared to be irreversible. Poor performance on IQ tests first noted in children with high blood lead levels appeared to persist into adolescence.

At the time of the accelerated phasedown of lead in gasoline, it is important to note that the evidence for lead's adverse effects on children was growing, although there was scientific disagreement about the danger to children from low-level exposures. No large reviews or syntheses of the scientific literature had yet appeared, but the weight of evidence for effects of low-level exposure to lead on children's cognitive function was increasing.

In response to the accumulating data on lead's hazardous potential, the Centers for Disease Control (CDC) steadily lowered its "concern threshold" for lead in the blood of children. In 1971, when reduction of lead in gasoline was first considered, blood lead levels below 60 μg/dl were considered safe. The concern threshold was lowered to 40 μg/dl in 1972, then to 30 μg/dl in 1975. The level has since been lowered again, first at about the time of the accelerated phasedown to 25 μg/dl, and finally to 10 μg/dl in 1991. Some scientific controversy has been engendered by specific recommendations for "thresholds of concern" and the shape of the dose-response relationship (is there a threshold below which lead does not affect IQ?), but today there is general agreement in the scientific community that lead has the potential to decrease cognitive function in children at levels of exposure below those causing clinical toxicity.

Better measurement and monitoring of lead in the human body enabled scientists to find more subtle effects of lead on adults. Epidemiological studies found associations between elevated blood lead levels and increased blood pressure in middle-aged white males (EPA, 1985). The relationship is important, because high blood pressure is linked to a number of serious cardiovascular events such as strokes and heart attacks. Although the effect of lead on blood pressure was relatively small, it appeared to occur over a wide range of blood lead levels.

As we will see, these changes during the 1970s and early 1980s in the scientific understanding of lead's hazards contributed to the risk assessment that ultimately drove the decision to speed up the removal of lead from the American gasoline supply.

The Emergence of the Catalytic Converter

In 1970 Congress passed amendments to the Clean Air Act that were designed to force Detroit's automobile industry to come up with new technology for reducing the emissions of hydrocarbons (from unburned fuel), carbon monoxide, and nitrogen oxides from cars. The new levels in these "technology forcing" regulations were set so low, corresponding to a more than 95 percent decrease in hydrocarbons for example, that they could not be achieved with minor adjustments to existing engine technology. Such strict levels were considered necessary for urban areas to meet new national ambient air quality standards (NAAQS) established under the Clean Air Act. These health-based standards were set by the EPA at levels that the agency judged would "protect the public health" with "an adequate margin of safety."

There were few compliance options available to the industry. Lean-burn engines were under development, but it was not clear that, even if successful, they could meet the EPA's stringent emission standards. Most attention was focused on catalytic converters, which could be added on to a car's exhaust system. The converters used a catalyst to complete the conversion of unburned and exhaust hydrocarbons to carbon dioxide and water. Although they resisted the idea at first, the U.S. auto manufacturers were eventually convinced that catalytic converters were effective and durable enough for general use. Indeed, the auto industry found that catalytic converters had additional benefits, since cars that had them could be tuned to optimize performance without concern about exhaust. For example, General Motors reported a 28 percent increase in fuel efficiency in vehicles with a catalytic converter. By the mid-1970s the catalytic converter became standard equipment on every new car as part of the EPA's plan to meet the NAAQS for ground-level ozone.

Lead versus the Catalyst

The introduction of the catalytic converter posed major problems for the oil and gasoline industry, however. Lead in gasoline "poisoned" the catalyst in the converter, quickly rendering it useless. This meant that a supply of unleaded gasoline, actually gas with less than about 0.07 grams of lead per gallon, would soon be necessary across the United States.

In January 1972 the EPA proposed two regulations relating to lead in gasoline (EPA, 1972). One provided for the general availability of unleaded gasoline by July 1974; the other proposed that the lead content of leaded gasoline be reduced to 1.25 grams per leaded gallon (gplg) by 1977. Specifically, the reduction of lead in leaded gasoline was justified by section 211 (1) of the Clean Air Act, which allows the agency to restrict the manufacture or sale of a fuel or fuel additive if "products of such fuel or additive will endanger the public health or welfare."

The EPA argued publicly that the regulations covering unleaded gasoline were intended primarily to ensure the success of the automobile catalytic converter in meeting federal clean air standards. But, EPA said, public health concerns about lead were what warranted the reduction in the lead content of leaded gasoline.

In January 1973 the unleaded gas rule was promulgated and the reduction of lead in leaded gas rule was reproposed (EPA, 1973a). Although the EPA had determined that reducing the amount of lead in gasoline would reduce a significant source of lead exposure for the population, especially for children, it was "difficult, if not impossible, to establish a precise level of airborne lead as an acceptable basis for a control strategy" (EPA, 1973b). However, since both urban adults and children were overexposed to lead, and because leaded gasoline was a source of environmental lead that could be significantly reduced, the EPA acted to restrict levels of lead in leaded gasoline.

The reproposed rule, rather than focusing on leaded gasoline per se, set maximum levels for lead content averaged over leaded and unleaded production by an individual refinery. The schedule was adjusted to make the targets for the early years easier to meet than they had been under the original proposal, but they would reach the same overall level by 1979. The final plan, which the EPA judged to be reasonable from both a health and an economic standpoint, was for an average between leaded and unleaded gasoline of 1.7 grams per gallon (gpg) in 1975, 1.4 gpg in 1976, 1.0 gpg in 1978, and 0.5 gpg in 1979. These target levels would be achieved by a combination of reduction in leaded fuel use (as a larger proportion of cars had catalytic converters) and reduction in the lead content of gasoline.

The EPA's rationale for reducing lead in gas was that this would reduce blood lead levels and the risks of lead poisoning. Lead poisoning, defined as the appearance of clinical symptoms (in contrast to later definitions, as we will see below) and associated with blood lead levels of at least 50μg/dl, was indeed occurring. The EPA also made note of, but did not hang its argument on, recent findings suggesting possible neurological effects on children at lead levels below those causing clinical symptoms.

The 1973 rule was not adopted without controversy. Most of the controversy, and technical challenges, to the proposal were based on the link between lead in gasoline and increased blood lead levels. Opponents of the rule suggested that gasoline lead was a small source of lead exposure, dwarfed by lead paint chips and lead in food. The EPA argued that there was evidence linking leaded gas to high lead levels in the air and dust, and that this was an exposure pathway "which can be readily and significantly reduced in comparison to these other sources."

The phasedown of lead in gasoline began according to the schedule proposed in 1973. The costs of meeting the average compliance targets differed for large and small refineries, and small refineries began to ask for relief. In January 1977 the EPA proposed a change in the lead phasedown targets, allowing small refiners (those with a capacity of less than 30,000 barrels of crude oil per day) a lead standard of 1.0 gpg averaged over a three-month period (EPA, 1977). Before the EPA could act, however, Congress wrote an amendment to the Clean Air Act in August 1977 that exempted small refineries from the old standard and set new targets based on capacity. The EPA wrote regulations, later codified, providing an exemption from the 0.5 gpg standard until October 1, 1982.

| Deregulators Target the Lead Rule

In January 1981 Ronald Reagan took office with a goal of reducing the regulatory burden on American business. The Task Force on Regulatory Relief, chaired by Vice President George Bush, was established by the Reagan administration with the mission of relieving industry of costly, irrelevant, or antiquated regulations. The group targeted a list of regulations perceived to be unnecessary and required that the appropriate agencies review the rules. In August 1981 the task force ordered the EPA, under the leadership of Anne Gorsuch, to consider permanently suspending the deadline for small refiners and to investigate the usefulness of the phasedown.

The task force knew that lead was on the way out. Since 1975 all new cars had come equipped with a lead-intolerant catalytic converter and were therefore supposed to be fueled with unleaded gasoline. The number of pre-1975 vehicles was diminishing each year as the vehicular fleet turned over. The big refiners, which supplied the majority of America's gasoline, had met the 0.5 gpg target in 1979. Thus, initially it was believed that market forces created by the requirement of the catalytic converter would be sufficient to complete the removal of lead in gasoline and that no further regulation of the small refiners would be necessary.

On February 18, 1982, the EPA requested comments on whether to relax or rescind the 0.5 gpg standard for leaded and unleaded gasoline combined. At the time of the announcement, unleaded gasoline represented 52 percent of the gasoline sold. At issue was whether the percentage of the vehicle fleet using unleaded fuels and the rate of fleet turnover were great enough that regulations would no longer be necessary. The agency indicated that it would consider "health effects, air quality impact, economic impact and energy impact." In the case that the standard was continued, the EPA also would examine different averaging schemes to provide flexibility for refiners.

In addition, the EPA indicated it would reconsider the requirements for small refiners. The agency proposed to suspend indefinitely the October 1, 1982 deadline for small refiners to meet the 0.5 gpg standard. Refineries processing no more than 50,000 barrels of crude oil per day and not owned by large refiners were adhering to lead limits of 0.8 to 2.65 gpg, depending on specific production levels at the time. The agency requested comments relating to adverse health effects, short-term financial savings, a standard for small refineries, the definition of a small refinery, intrarefiner or interrefiner averaging, and standards for the final product of imported gasoline. Until this official review of the lead regulations, the Task Force on Regulatory Relief was not aware of several confounding factors that were sabotaging the success of the leaded gas phasedown.

The most serious concern of the EPA was misfueling, the use of leaded gasoline in vehicles equipped with catalytic converters. Although all post-1975 vehicles were designated unleaded, about 15 percent of these vehicles were misfueled either accidentally or intentionally by drivers interested in boosting octane or saving a few cents per gallon on gasoline. Because leaded gasoline poisons a catalytic converter, this misfueling resulted in an increase in hydrocarbons and nitrogen oxides, thus exacerbating the problems of surface ozone and Clean Air Act violations. The lead also damaged the oxygen sensor, which monitored the state of the catalytic converter in newer cars, reducing fuel efficiency significantly. Misfueling increased the demand and use of leaded gasoline beyond the agency's projections, thereby increasing lead emissions. Plus, leaded gasoline poisoned the catalytic converter without affecting driving performance. Thus a misfueled vehicle was capable of unusually high lead and primary air pollutant emissions without the driver's even realizing it.

There was a great deal of consternation in the oil refining industry at this time. Many large refiners had made large capital investments to comply with the lead phasedown, and they were unhappy with the idea that small refiners might be getting special treatment. Other large refineries had not yet made the investment and hoped to catch a break along with the small refiners. The small refiners, for their part, felt that the regulations, as they stood, would put them out of business. In addition, new types of facilities, known as gasoline blenders, were beginning to crop up to exploit the loophole created by the extended deadline for the small refiners. Blenders purchased low-quality, cheap gasoline and added lead to boost the octane and upgrade the gas to market specifications. Although blenders did not actually refine crude oil, their production size enabled them to be regulated as small refiners. These facilities were jeopardizing the nation's efforts to reduce the total quantity of lead in gasoline.

In this charged atmosphere, a rather large problem developed for Anne Gorsuch. Some large refiners and members of Congress complained that she was giving preferential treatment to small refiners. Representative Toby Moffett (D-Conn.), chairman of the House Government Operations Subcommittee on Environment, Energy, and Natural Resources, accused Gorsuch of promising a small refiner from New Mexico that it would be allowed to exceed the lead standard applicable to small refiners. The decision to grant special privileges to the Thriftway Company of Farmington, New Mexico, showed that Gorsuch would easily "grant amnesty to parties fortunate enough to be able to meet with [her] in private," which "constitutes a serious breach" of her "oath of office and the code of ethics for government employees." Moffett also requested that the inspector general of the EPA investigate the case. The inspector general found that Gorsuch indeed promised Thriftway it wouldn't be penalized for violating lead standards. The company had claimed that it would otherwise be forced out of business. The furor over Gorsuch's meetings with small refiners, along with many other factors, lead to her resignation as EPA administrator.

On August 27, 1982, six months after its request for comments, the EPA announced that the 0.5 gpg standard would not be relaxed or rescinded. The agency had determined that 1) relaxing or rescinding the standard would result in an increase in the amount of lead released into the atmosphere, 2) environmental lead exposure remained a national health issue, and 3) there was no new information that would give the agency reason to believe that continuing the regulation of lead in gasoline would not be necessary (EPA, 1982a). The Task Force on Regulatory Relief, including staff at the Office of Management and Budget, agreed with the agency's decision. Since this was a decision *not* to act, it did not come under President Reagan's Executive Order 12291, which required certification that the benefits of an action would outweigh the costs.

After looking closely at the question, the EPA came back and affirmatively *proposed* that the 0.5 gpg standard for the average lead content in both leaded and unleaded gasoline produced by each refinery be replaced with a two-tiered standard on leaded gasoline only (EPA, 1982a). There was concern in the agency that a standard based on the level of lead in all gasoline might not reduce lead use and emissions, because the amount of lead in leaded gas could increase as unleaded gas sales increased. Large refineries would have a standard of 1.10 gplg, while 2.50 gplg would be the standard for small refineries, though this proposal would change in 1984. The EPA also stipulated that the average lead content of imported leaded gasoline sold or offered for sale must be less than 1.10 gplg. In this same notice, the EPA proposed a trading program in which two or more refineries owned by the same refiner would be able to average lead content quarterly in order to comply. EPA also tightened the definition of small refinery in order to reduce the number of facilities eligible for less stringent standards.

At the end of 1982 all of the confusion about lead levels was settled. On October 29, 1982, the EPA promulgated revised regulations (EPA, 1982b) in which the basis of the standard was changed from total gasoline (leaded and unleaded) to a standard for lead in leaded gasoline only, set at 1.10 gplg. Large refineries and importers were subject to a lead-content standard of 1.10 gplg, effective November 1, 1982, and small refineries would be subject to the same standard of 1.10 gplg, but effective eight months later. Trading of lead use across refineries was allowed such that the industry as a whole would meet the standard. In addition, the definition of small refiners was significantly changed, resulting in a reduction in the number of facilities able to take advantage of the interim standard for small refiners.

The result, therefore, was that the 0.50 gplg standard for the average of all gasoline was replaced by a 1.10 gplg cap on leaded gasoline only. Small refiners were granted until July 1, 1983, to meet this standard. The amount of lead emitted by American automobiles would continue to decline as more and more cars designed to run on lead-free gas were put on the road.

| Origins of the Accelerated Phaseout

In the spring of 1983, William Ruckelshaus was recruited back to the EPA, replacing Anne Gorsuch as administrator. Ruckelshaus appointed Alvin Alm, a professional with strong commitments to using risk analysis in regulatory decision making, as deputy administrator. At this time the issue of lead in gasoline had drifted into the background. Most of the parties involved had accepted the slightly stricter standards and the trading program. The phasedown was apparently successful, and blood lead levels as well as ambient lead concentrations were decreasing.

So what was it that initiated the EPA's efforts to accelerate the already existing lead phasedown? The idea can be traced through the agency to Alm, its new deputy administrator. After a meeting with a representative from the ethanol industry on an issue unrelated to lead, the representative, on his way out, asked Alm why the agency would not consider phasing lead out of gasoline completely. There were so many problems associated with lead, such as misfueling, vehicle damage, and direct health effects — why not just get rid of it once and for all? Alm was intrigued with the idea. It had the possibility of extremely significant benefits and seemed a perfect opportunity for serious analysis. He believed that a structured analysis could identify and quantify the social costs and benefits of removing lead entirely from gasoline in comparison to the current phasedown (Alm, 1994). In addition, the political situation was also conducive to the idea, for the Reagan administration and the EPA were looking to buttress their pro-environment image.

Shortly thereafter, Alm discussed the idea with Milton Russell, the assistant administrator of the Office of Policy, Planning, and Evaluation. They posed to Joel Schwartz, a staff member, the following question: "Given that lead is harmful to human health and that ozone non-attainment is a serious agency problem, would it be worthwhile to accelerate the phasedown of lead in gasoline?" Schwartz saw this as an opportunity to apply the tools of cost-benefit analysis to an important problem (Schwartz, 1994). He began his investigations with some "back of the envelope" calculations. The results made Alm and Russell immediately ask for a more formal analysis. A team with expertise in engineering and economics was assembled to work with Schwartz. The project members were Schwartz and career EPA analysts Hugh Pitcher, Ronnie Levin, and Bart Ostro.

It is important to point out that the project started as an analytical endeavor inside the EPA. There were no statutory requirements or deadlines. There was no political influence. Neither Alm nor Russell was under pressure from external groups to accelerate the phasedown. Furthermore, once the project was started, it was kept very quiet. Even people in other EPA offices were unaware that the analysis was under way. The regulation of lead in gasoline was a politically sensitive issue after the battle in the Gorsuch administration, and Alm, Russell, Schwartz, and the rest of the analysis group wanted to keep a low profile. However, it is likely that all involved felt a political motivation to help the EPA regain its pro-environmental credentials. This situation was a bit unusual, in that the analysis was actually being done in the EPA's policy office, which in most situations looked over the shoulder of other program offices conducting analyses.

Russell grew concerned that the analysis was not receiving the continuous checks and balances typical for openly discussed projects in the agency. He therefore moved Albert Nichols, who had joined the agency from Harvard University's Kennedy School of Government, from his position as special assistant to Russell to that of division director of economic analysis (above Schwartz), to serve as an objective reviewer and contributor to the analysis and regulations.

Alm, Russell, and the analysis group did not start with a particular regulatory decision in mind but instead simply asked what would be the costs and benefits of going further with the phasedown of lead in gasoline. In March 1984 Schwartz and his colleagues released their draft final report, *Costs and Benefits of Reducing Lead in Gasoline.* The preliminary analysis compared the costs and benefits, in 1988, of three scenarios: 1) banning lead entirely from gasoline by 1988; 2) allowing a low-lead fuel; and 3) maintaining the present 1.10 gplg standard. The EPA staff decided that the analysis did, in fact, justify more aggressive regulation of the lead content in gasoline. It is crucial to note that, in this case, the proposed regulation was driven by an analysis — that the analysis had *not* been done to support a rule that had already been proposed.

The study found that although the first option, a ban on lead, would abate the problems associated with misfueling and the direct health effects of lead emissions, it carried notable costs. First, some pre-1971 vehicles actually required lead as a valve-seat lubricant and would therefore suffer damage if fueled with unleaded gasoline. Second, lead is the least expensive octane booster available, and its removal would require refiners to further process crude oil, entailing a more costly method of octane enhancement. However, the study estimated that in total this would increase the refineries' costs by only 1 percent. A third concern was that a switch to unleaded gasoline could lead to regional shortages or an increased dependence on foreign oil. Methods of making high-octane lead-free gasoline require more oil per gallon of gas than when lead is used to boost octane. With the oil shortages of the 1970s still fresh in their minds, the analysis group conducted extensive modeling exercises to evaluate the regional and national impacts on gas and oil consumption of a switch to lead-free gasoline.

The second scenario, the low-lead option, allowed 0.10 grams of lead per gallon. This was considered enough lead to protect certain older vehicles from valve-seat recession while still reducing airborne pollution. The third scenario, maintaining the status quo, served as a base case from which to draw comparisons.

Benefits analyzed included improved vehicle maintenance and fuel economy benefits as well as environmental and health benefits. The largest monetized benefits were found to be the savings from vehicle operations and maintenance. Lead and the scavengers used to remove lead from the engine after combustion are extremely corrosive to an automobile's engine and exhaust system. Extensive analysis of U.S. Army and Postal Service experience indicated that a ban on or reduction of lead in gasoline would decrease the number of tune-ups, exhaust system replacements, and oil changes needed by vehicles. In fact, the equivalent of three to four cents per gallon of gasoline could be saved in maintenance costs if vehicle owners who misfueled switched to unleaded gasoline. Fuel economy benefits arise because misfueled cars, in which lead fouls the oxygen sensors of the pollution control system, use more fuel. The benefits simply from increased fuel economy in cars were estimated to offset 15 to 20 percent of the costs of the regulation.

The health benefits analyzed included the benefits of reducing conventional pollutant emissions and lead emissions. Lead poisons catalytic converters that are designed to remove hydrocarbons, carbon monoxide, and nitrogen oxides, the so-called conventional pollutants. Although all post-1975 vehicles were equipped with catalytic converters, 15 percent of those vehicles were being misfueled with leaded gasoline. The estimates of contaminant reduction on which the monetizations were based assumed that misfueling would be completely eliminated with both the low-lead and no-lead options. This was believed to be a reasonable assumption because as lead-free gas became the norm it would become less expensive than regular leaded gas, removing the strongest incentive to misfuel.

Banning or limiting lead would also decrease the direct health effects associated with conventional pollutants, as well as their contribution to ozone formation. The excess emissions of conventional pollutants due to misfueling were estimated and used to value the reduction of primary pollutants, according to several methods. Scientific uncertainty, insufficient data, or difficulties in monetary valuation prevented a complete monetization of the health benefits of eliminating misfueling. For example, the effects of ozone on ecosystems, the effects of chronic ozone exposure on human health, and carbon monoxide emissions were not quantified but addressed qualitatively. The monetary valuations for reduced health risks from conventional pollutants may therefore have been underestimated significantly.

The analysis valued the benefits of lowering blood lead levels by determining the number of children whose blood lead levels would be brought below the 30 μg/dl standard of lead toxicity. Several recent publications had correlated lead emissions from automobiles with the blood lead levels of children. For example, the second National Health and Nutrition Examination Survey had found a strong statistical correlation between blood lead levels and the amount of lead in gasoline. Schwartz used this relationship in a forecasting model. Only the benefits of reduced medical care for severe lead toxicity and reduced cognitive damage were monetized. The latter category included the costs of additional education to compensate for IQ deficits.

Many benefit categories could not be quantified. The draft version mentioned but did not address the adverse effects of lead on adults, including kidney damage and anemia; effects on fetuses and infants less than six months old; behavioral problems in children that resulted in decreased attention; lost work and leisure time by family members of affected children; and long-term social costs of a learning disabled workforce. Moreover, the document discussed but did not monetize the public health benefits of reducing children's blood lead levels below the 30 μg/dl level of concern. Improved monitoring and measuring had enabled scientists to observe pathophysiological responses at levels previously thought to be safe, prompting the CDC to consider lowering the 30 μg/dl standard for lead toxicity.

The annual automobile maintenance savings estimates were significant, \$660 million for the low-lead option and \$755 million for the no-lead alternative. A large source of benefits arose from reduced corrosion of exhaust systems and engines. Scavengers added to gasoline to reduce lead deposits made exhaust from leaded gasoline much more acidic than that from unleaded gas, increasing costs of maintaining exhaust systems. Unleaded gas would also extend the life of spark plugs and extend oil change intervals. These maintenance savings alone offset the total monetized costs of the rule. The low-lead option had benefits exceeding costs by \$64 million; for the no-lead plan the difference was \$157 million.

The preliminary analysis represented an enormous effort by Schwartz and his colleagues during the final months of 1983. Before its March 1984 publication, the report was peer reviewed and fine-tuned in an unofficial meeting with OMB staff. To meet the standards of the 1981 Executive Order (EO) 12291, which requires a regulatory impact analysis for any "major regulation," the authors made slight alterations to the existing report to satisfy format requirements. The EO 12291 authorizes the OMB to override EPA decisions regarding "major regulations" if the potential benefits do not outweigh the potential costs. Many considered the requirements of EO 12291 a way to delay and prevent regulation. The Reagan administration, on the other hand, claimed that it only blocked or delayed poor regulations, that rules that made sense would be approved. The preliminary regulatory impact analysis and the proposed rule on lead in gasoline were submitted to the OMB before their publication in the *Federal Register* on August 2, 1984.

A Victory for Analysis

In spite of the what was widely considered an antiregulatory attitude at the OMB, the office approved the acceleration of the lead phasedown in gasoline, thanks to the quality of the analysis. In fact, this was the kind of analysis that the OMB dreamed about. It was a serious cost-benefit study. The numbers were solid. Where possible, benefits were rigorously monetized, which the OMB staff appreciated. According to Chris DeMuth of the OMB, the office was opposed to only senseless regulations; the analysis demonstrated that the accelerated phasedown was reasonable, and so the OMB supported it (DeMuth, 1994). According to Milton Russell, the regulatory impact analysis was a perfect example of the type of regulation and supporting analysis that the White House and the OMB just could not oppose. In addition, the EPA would be able to take a high-profile proregulatory stance that would help it to regain some of the credibility it had lost in the past few years.

On August 2, 1984, the EPA proposed to accelerate the phasedown of lead in gasoline by revising both the short-term and long-term standards. With respect to the short-term, the agency proposed to replace the lead content standard of 1.10 grams of lead per gallon in leaded gasoline with a stricter limit of 0.10 gplg in total gasoline produced, effective January 1, 1986. The notice of proposed rule making (NPRM) also contained an alternative phasedown schedule in the case that comments proved the proposed date infeasible. With respect to the long-term schedule, two approaches aimed at the complete removal of lead from gasoline were proposed: 1) a ban on lead usage in gasoline by 1995, and 2) no additional regulatory action, based on the assumption that the turnover of the fleet and use of modern engines would eliminate the use of lead.

According to the EPA, continued problems with misfueling and new information regarding the adverse health effects of lead motivated the agency to propose the stricter standards. The amount of lead used in 1983 was 51.83 billion grams (about 58,000 tons), which was 10.4 percent higher than the 1982 EPA predictions. The agency explained that misfueling, as well as higher-than-expected total gasoline demand, longer retention of older vehicles, and the small refiners' standard, meant that the demand for leaded gasoline remained high, rendering the ongoing phasedown less effective.

The agency reported that new information since the 1982 rule had reinforced concern about lead toxicity, especially that associated with low blood lead levels. New studies also strengthened the EPA's previous conclusion that a relationship exists between gasoline lead usage and blood lead levels. The notice of proposed rulemaking stated that the "EPA used the quantitative results of these new analyses in examining various regulatory alternatives and impacts, such as the impact on blood lead levels of children as gasoline lead levels varied, and EPA expects to consider these results in formulating any final rule."

The August 2 proposed rule relied heavily on the analysis by Schwartz, Pitcher, Levin, Ostro, and Nichols to quantify the impact of the regulation on total lead use, the economy, human health, air quality, and the use of other fuel additives. The major consequence of the rule was stated to be decreased human exposure to environmental lead, especially by the high-risk group, young children. The proposed 0.10 standard was predicted to prevent 280,000 cases of high blood lead levels in children (those measuring more than 30 μg/dl) and 9.6 million incidences of blood lead levels above 15 μg/dl.

The economic analysis in the preliminary regulatory impact report in the proposed rule estimated that although meeting the 0.10 gplg standard would cost the refining industry $3.4 billion from 1986 to 1992, the overall benefits, during this same period, would amount to $10.7 billion.

| Reaction to the Proposal

Public hearings on the proposed rule, held on August 30 and 31, 1984, generated more than 1,500 written comments and oral testimony from sixty witnesses. These responses came from a diverse group, ranging from refiners, blenders, representatives of the lead industry, and gasoline marketers, to physicians, environmentalists, and health groups. Predictably, major criticism of the document and proposed regulations came from the Lead Industries Association and Ethyl Corporation. Larry Blanchard of Ethyl had earlier argued that if the agency were concerned with misfueling then it should attack that problem directly and that there was "no sound evidence linking lead in gasoline and lowering IQ in children" (*International Petrochemical Report,* 1984). With respect to the Schwartz report, Blanchard accused the EPA of "intellectual fraud" by misinterpreting the data that correlated blood lead levels with lead in gasoline (*Environment Reporter,* 1984).

The oil refiners, who comprised the regulated industry, were divided on the issue. Although some refiners strongly opposed the standard, most accepted the lead limit as necessary to protect public health but were concerned that the deadline might not be feasible. The refiners that had made capital investments and had sufficient refining capacity to meet the new standards were in agreement with the proposed deadline and even lobbied for one earlier than January 1, 1986. They asserted that any delay in the effective date of the standard would penalize companies that had made capital improvements and reward those that had been "dragging their feet." Some companies claimed that industrywide compliance would bring about gasoline shortages and economic "hardship." Companies that had already made the necessary capital improvements stood to make a lot of money from an early deadline (Dicorcia, 1994).

In response to the concerns voiced about the economic and technological feasibility of the proposed deadline, Schwartz and his coworkers conducted, and included in their final regulatory impact analysis, extensive analyses showing that even under conservative assumptions, the industry had sufficient capacity to meet the standards, and that only in the case of several pessimistic assumptions occurring simultaneously would the standard be infeasible. In addition, a proposed "banking" program would provide flexibility by allowing refiners using less than the allowable lead limit in 1985 to bank extra lead usage rights to use later toward meeting the stricter 1986 and 1987 standards.

The EPA also responded to the concerns of other affected parties, including the lead industry, antique car owners, boaters, motorcyclists, truck fleet operators, farm groups, and some engine manufacturers. These groups were worried that a 0.10 gplg lead limit would not be sufficient to protect engines from valve-seat recession. Significant evidence presented by others, including studies and real-world experience, along with the project group's own analysis, revealed that valve damage at the new lead level would be minimal, enabling the agency to stick to the proposed 0.10 gplg standard.

Based on the public comments and even more recent scientific information, the EPA decided to act on the short-term and long-term standards separately. On March 7, 1985, the EPA published two notices. With respect to the short-term standard, the agency promulgated the 0.10 gplg standard for lead in leaded gasoline effective January 1, 1986, with an interim standard of 0.50 gplg effective July 1, 1985. This rule also eliminated refinery averaging as of January 1, 1986. In the second notice, the EPA reopened the comment period on the question of an eventual ban on lead in gasoline. The EPA requested comments on new scientific and technical information that had been developed or received since the original August 1984 proposal. According to the notice, the new information suggested that the health consequences of lead in gasoline might be more significant than previously thought, that misfueling would continue to be a national problem, and that lead was not so important as a valve lubricant as originally believed.

With respect to human health, new findings had suggested a relationship between blood lead levels and blood pressure in adult men. The agency was extremely concerned about the serious public health implications of these findings, because increased blood pressure is correlated with risk of heart attack, stroke, and premature death. Based on data from one of the studies the final regulatory impact analysis estimated that lead in gasoline may result in more than 1 million cases of hypertension and more than 5,000 deaths from diseases related to hypertension, such as heart attacks or strokes. In addition, the CDC had lowered the level of concern for lead in blood from 30 μg/dl to 25 μg/dl, effectively tripling the number of children considered at risk.

The CDC commented that because of difficulty in establishing a threshold for the effects of lead on children's cognitive function, the lower limit was still not necessarily "safe."

The EPA's *Final Regulatory Impact Analysis,* issued in February 1985, supported the reduction of lead in gasoline from 1.10 to 0.50 gplg on July 1, 1985, and to 0.10 gplg on January 4, 1986 (EPA, 1985). The final analysis evaluated the costs and benefits of alternative lead levels, focusing on 1986, the first year in which the standard would be in effect. Clearly the lowest lead level, 0.10 gplg, achieved the greatest net benefit, with the costs of the regulation being far outweighed by the benefits estimated (Table 2.1). The degree of misfueling, which leads to poisoning of the catalytic converter and increased levels of conventional pollutants, was modeled at several levels. The report stated that at the 0.10 gplg standard, the price of leaded and unleaded gas would eventually converge and most misfueling would end, and that therefore benefit values calculated with no misfueling were appropriate. The values of the costs and benefits for the final rule, projected for the years 1985 to 1987 (in millions of 1983 dollars) were: costs, $1,130; benefits, $16,095; net benefits, $14,965. Excluding the benefits from reduced blood pressure, the net benefits were still $2,924 million. The analysis used a discount rate of 10 percent for future benefits; lower discount rates would have made the benefits even larger.

Table 2.1 reflects several changes that were made in response to

TABLE 2.1 | Costs and benefits of different levels of lead in leaded gasoline for 1986, assuming no misfueling (millions of 1983 dollars)

	LEAD LEVEL (GPLG)				
MONETIZED BENEFIT	0.50	0.40	0.30	0.20	0.10
Children's health effects	$ 466	$ 504	$ 539	$ 571	$ 602
Adult blood pressure	4,018	4,483	4,955	5,436	5,927
Conventional pollutants	278	278	278	278	278
Vehicle maintenance	517	608	706	808	933
Fuel economy	119	119	128	136	190
Total refining costs	243	305	386	472	607
Net benefits	5,155	5,687	6,220	6,757	7,323
Net benefits excluding blood pressure	1,137	1,204	1,265	1,321	1,396

Source: EPA, 1985, table VIII-3a.

comments to the draft regulatory impact analysis and new information. Estimates of total refining costs to meet a 0.10 gplg standard were raised from $575 million to $607 million. Estimates of benefits of reductions in conventional pollutants went down slightly, maintenance benefit estimates increased, fuel economy benefits decreased. The final regulatory impact analysis considered that reductions in lead in gasoline would lead to increased levels of aromatic compounds, including benzene. However, due to decreased misfueling, leading to better catalytic converter function, and use of evaporation control devices on new cars, the net risks from benzene were actually predicted to decrease. The benefits of the estimated decrease of 4.4 leukemia deaths per year were included in the final cost-benefit analysis. The decrease in emissions of ethylene dibromide, a possible carcinogen added to leaded gasoline as a scavenger, was noted as a benefit of the phasedown but not quantified. None of these changes had a significant effect on the relationship of benefits to costs. The biggest change was the inclusion of benefits due to reductions in adult blood pressure, which had not been considered in the draft analysis.

The report stated that the *"EPA's primary objective in promulgating this regulation is to minimize the adverse health and environmental effects of lead in gasoline."* The health and environmental benefits relied on to garner support for the regulation were decreased ill effects on children's health and cognition and the reduction in pollutants due to misfueling. Based on a suggestion from assistant administrator for Research and Development Bernard Goldstein, the EPA chose not to rely on the blood pressure health benefits "until there has been a greater opportunity for scientific review and public comment," although the monetized health benefits, using a relatively conservative value of $1 million per life, were the largest of all considered. The additional maintenance and fuel economy savings were also considerable, the vehicle maintenance savings alone outweighing the estimated costs of the regulation.

| Conclusions

It is often stated that quantitative analyses of risks, costs, and benefits by regulatory agencies serve only to delay, weaken, or obstruct regulation. Contrary to this skeptical view, this chapter illustrates a situation in which a study of benefits and costs convinced an antiregulatory administration to tighten and then accelerate regulations to protect human health and the environment.

As a result of this rigorous analysis, the United States emerged as a leader in removing lead from gasoline. Many other countries, including the countries of the European Union and the Pacific Rim, have lagged far behind. Even in 1996 the unleaded share of the E.U. market, excluding diesel, was only 67 percent. And in four countries — Greece, Spain, Italy, and Portugal — consumption of leaded gasoline still exceeds unleaded (Dow-Jones News Service, 1996). Now, often based on the U.S. analysis or similar work, most countries in the world are removing lead from the gasoline they use.

Several factors may have contributed to the decision to accelerate the phasedown of leaded gasoline in this country. For one, the decision was not all or nothing; instead it "fueled a moving train," simply accelerating the pace at which lead was being removed from gasoline. Clearly the ability to rely on the "hard" monetary benefits from lower maintenance and fuel economy reduced the impact of arguments about the way in which "soft" benefit estimates, such as costs of remedial education, had been developed. In addition, the decision allowed the Reagan administration to demonstrate a commitment to environmental protection, deflecting some of the criticism arising from its deregulation efforts. The use of cost-benefit analysis also gave the Reagan administration the ammunition it needed to stand up to its constituency, the business community, and declare that the regulation was needed. The quantification of benefits to children may have also made it difficult for the lead industry to fight very hard without appearing heartless.

Although the issue of lead in gasoline had had a very high profile during Anne Gorsuch's tenure as EPA administrator, the accelerated phasedown was not a big story in the press. Looking back, several people closely involved with the process expressed surprise at the fact that members of the press and environmental organizations took credit for influencing the regulation (Alm, 1994; Russell, 1994). In fact, it may have been the quality of the supporting analysis that allowed the regulations to move through the regulatory process with remarkable speed, only seven months from the time of the proposed acceleration to the promulgation of the stricter standard. In this case, the fact that the net benefits were so overwhelming, even without consideration of blood pressure effects, may have meant that the industry saw no point in arguing the technical merits of the analysis. There was virtually no way that it could make the costs appear anywhere close to the benefits. If the ratio had been closer, it is possible that the pace of regulation may have been considerably slower.

The accelerated phasedown of lead in gasoline also gave a considerable boost to the morale of EPA staff. According to Milton Russell, agency personnel often took a "ho-hum" view of many regulations, but the lead phasedown was universally hailed as a major public health advance. Many believed that this was the greatest single action the EPA had taken and, in fact, see few opportunities to match it in the future (Russell, 1995).

This case provides a clear example of how careful analysis can identify opportunities for protecting the public health. Consideration of both costs and benefits sold the more stringent regulatory program to the EPA, the administration, and the affected industries.

| Sources

Alm, A., 1994. Interview August 10.

DeMuth, C., 1994. Interview August 25.

Dicorcia, E., 1994. Interview with Ed Dicorcia, Vice President of Refining, Exxon, August 11.

Dow-Jones Business News, "E. U. Sustainable Transport Policy Harmed by Fuel Use Growth" (September 9, 1996).

Bureau of National Affairs, September 7, 1984. "Opponents of Lead Phasedown Proposal Say Health Effects Data Misinterpreted," *Environmental Reporter,* 15, no. 19.

Environmental Protection Agency (EPA), 1972. "Regulation of Fuels and Fuel Additives: Lead and Phosphorous Additives in Motor Vehicle Gasoline," *Federal Register* 37:3882.

———, 1973a. "Regulation of Fuels and Fuel Additives," *Federal Register* 38:1254.

———, 1973b. "Regulation of Fuels and Fuel Additives: Control of Lead Additives in Gasoline," *Federal Register* 38:33734.

———, 1977. "Regulation of Fuels and Fuel Additives," *Federal Register* 42:3188.

———, 1982a. "Regulation of Fuels and Fuel Additives," *Federal Register* 47:38070.

———, 1982b. "Regulation of Fuels and Fuel Additives," *Federal Register* 47:49322.

———, 1985. *Costs and Benefits of Reducing Lead in Gasoline: Final Regulatory Impact Analysis* (Washington, D.C.: Government Printing Office).

Goyer, R. A., 1991. "Toxic Effects of Metals," in M. O. Amdur, J. Doull, and C. D.

Klaassen (eds.), *Casarett and Doull's Toxicology: The Basic Science of Poisons* (New York: Pergamon Press).

International Petrochemical Report 3, no. 11 (March 13, 1984).

Needleman, H. L. (ed.), 1992. *Human Lead Exposure* (Ann Arbor, Mich.: CRC Press).

Needleman, H. L., C. Gunnoe, A. Leviton, et al., 1979. "Deficits in Psychologic and Classroom Performance of Children with Elevated Dentine Lead Levels," *New England Journal of Medicine* 300:689–695.

Rosner, D., and G. Markowitz, 1985. "A 'Gift of God'? The Public Health Controversy over Leaded Gasoline during the 1920s," *American Journal of Public Health* 75:344–352.

Rummo, J. H., D. K. Routh, N. J. Rummo, and J. F. Brown, 1979. "Behavioral and Neurological Effects of Symptomatic and Asymptomatic Lead Exposure in Children," *Archives of Environmental Health* 34:120–124.

Russell, M., 1995. Interview with Milton Russell, October 5.

———, 1994. Interview with Milton Russell, August 2.

Schwartz, J., 1994. Interview with Joel Schwartz, June 29.

Smith, M. A., L. D. Grant, and A. I. Sors (eds.), 1989. *Lead Exposure and Child Development: An International Assessment* (Hingham, Mass.: Kluwer Academic Publishers).

Yule, W., R. Lansdown, I. B. Miller, and M. A. Ubanowicz, 1981. "The Relationship between Blood Lead Concentrations, Intelligence, and Attainment in a School Population: A Pilot Study," *Developmental Medicine and Child Neurology* 23:567–576.

THREE

Protecting the Ozone Layer

James K. Hammitt
Kimberly M. Thompson

Since commercial introduction in the early 1930s, a single family of chemicals has revolutionized the American way of life. While chlorofluorocarbons (CFCs) were first used as refrigerants, they have since been used as solvents, blowing agents for foams and insulation, and propellants for aerosols. A closely related class of compounds, halons, has the unique ability to smother fires without leaving a residue and without producing toxic effects. By the early 1980s, CFCs and halons had permeated many aspects of daily life. For example, CFCs used as refrigerants and in the insulating material in refrigerators kept food fresh in homes, markets, and during transport; CFCs used in air conditioners and building insulation improved comfort in homes, workplaces, and vehicles; and halons used in fire extinguishers lessened the impact of fire damage. In 1988, as international regulation of CFCs and halons began, more than 10,000 U.S. companies employing 700,000 workers depended on CFCs to produce over $28 billion in goods and services per year, and the value of installed equipment that required CFCs exceeded $135 billion (Cogan, 1988). In that year, the $750 million U.S. industry led the world in CFC and halon production, and the United States accounted for about 30 percent of world consumption of these compounds, with annual per capita consumption exceeding 1 kilogram (Cogan, 1988).

In spite of our dependence on these substances, we now face the challenge of living without them, due to global concern about the risks they pose. In the United States and other industrialized countries, production of the primary CFCs came to an end in 1995. While the nonflammability, nonreactivity, and low toxicity of these substances make them "miracle chemicals" for a wide variety of applications, the same properties also give them the ability to modify the earth's atmosphere. Specifically, these substances destroy stratospheric ozone, which protects the earth from harmful ultraviolet radiation, and they contribute to global warming.

The case of CFCs and halons is unique. It inspired the first global agreement to address an environmental problem — the 1987 Montreal Protocol on Substances that Deplete the Ozone Layer. The success or failure of the agreement will depend on continued international cooperation and ultimately on the responses of producers and consumers worldwide. In this chapter, we describe the historical factors leading to the Montreal Protocol and the related changes in industrial activity. We evaluate the role of risk assessment in producing these changes. Consistent with the other cases in this book, we direct our attention to developments in the United States, although other nations, particularly Canada and several nations in Scandinavia and Europe, introduced national regulations at about the same time as the United States and played significant roles in the process leading to the current international regulatory structure.

We begin with a description of the development of the industries that used CFCs or halons in their production processes. Subsequently, we review the unfolding scientific developments that led to concern about CFCs, the accompanying risk analyses and government regulations, and the industrial and public response. In presenting this history, we find it useful to distinguish three periods:

- A period of *initial concern* (1970 to 1978), during which early research on the effects of air- and spacecraft on the stratosphere and on the environmental consequences of CFC use led to the 1974 discovery that CFCs could deplete stratospheric ozone; the public, industrial, and government response led to reductions in use and a ban on using CFCs in aerosol containers.
- A period of *watchful waiting* (1978 to 1985), in which increasing scientific uncertainty, economic slowdown, and an anti-regulatory administration produced little change.
- A period of *increased concern and international action* (1985 to

the present), including discovery of the Antarctic ozone hole, observation of decreases in global ozone, and the negotiation and implementation of the Montreal Protocol and subsequent amendments.

Although our focus is on the period leading up to the 1987 signing of the Montreal Protocol, we provide a brief summary of subsequent developments, including amendments to the protocol that increased its stringency and more recent industrial changes. In the final part of the chapter, we analyze the extent to which risk analysis contributed to regulation and industrial change, and assess the extent to which these changes can be viewed as improving environmental quality.

| Development of Key Industries

Several key industries began using CFCs or halons at different times to meet new consumer needs.[1] The industries use the compounds in dramatically different ways, in some cases resulting in their quick emission into the atmosphere and in others incorporating them into products from which they are released slowly, over decades. Differences in the amounts and ways the industries use these compounds influence the relative abilities of the industries to respond to risk concerns.

Refrigeration

The use of CFCs began with the refrigeration industry. In the refrigeration cycle, refrigerant vapor is compressed then liquefied by cooling via air or water. The liquid refrigerant is vaporized when it absorbs heat at constant pressure in an evaporator cooling coil. The chilled air circulates through an insulated box to create a refrigerator. Similar principles are used for air conditioning. The cost of the refrigerant typically represents a relatively small part of the overall cost of the refrigeration system.

In 1928 Thomas Midgley, Jr., was charged with the task of creating a superior refrigerant, and he synthesized the first CFC for the Frigidaire Division of General Motors. During the next two years, Midgley synthesized a number of different CFCs and evaluated their use as refrigerants in domestic refrigerators (Roan, 1989). In 1930 Frigidaire selected dichlorodifluoromethane (CFC-12) as the optimal

[1] This section relies on Cogan (1988) and Hammitt et al. (1986).

refrigerant, because it vaporized at relatively low temperatures and pressures and was low in toxicity, nonreactive, and nonflammable. In addition, CFC-12 appeared to be relatively inexpensive to produce. General Motors formed a joint venture with E. I. Du Pont de Nemours and Company (Du Pont) to produce CFC-12, and in 1931 Du Pont produced approximately 500,000 kg of CFC-12 under the trade name Freon. In 1933 the first Frigidaire domestic refrigerators using CFC-12 were on the market.[2]

In 1934 Du Pont began production of trichlorofluoromethane (CFC-11) for use in centrifugal chillers (used for air conditioning large buildings) and as an industrial coolant. By the 1940s refrigerators using CFC-12 dominated the domestic refrigeration market. In 1945 production of CFC-12 grew to 20 million kg per year as the production of domestic refrigerators not using CFC-12 ceased. During the 1940s and 1950s the use of CFC-12 and other CFCs for refrigeration quickly expanded into commercial refrigeration systems like the reciprocating systems used in supermarkets. Sales growth of reciprocating systems peaked in 1966 at almost 10,000 units. CFC refrigerants were also used to cool the refrigeration compartments in trucks and rail cars that transported food to markets. Sales of centrifugal chillers that used CFC-12 for air conditioning grew dramatically between the late 1940s and the mid-1960s as they were built into offices, schools, hospitals, and shopping centers. Living standards have changed so dramatically in the past sixty years that now more than 70 percent of the new homes built in the United States include central air conditioning (using HCFC-22[3]), with the percentage exceeding 90 percent in southern states. More important, the amount of CFCs used in the residential market is estimated to be less than 10 percent of the amount used in the commercial market.

The large quantities of CFCs used as working fluids in stationary refrigerators and air conditioners are swamped by the amount used for mobile air conditioning. Although air conditioners began as a luxury

[2] The CFC naming system was also introduced at this time. Each name starts with "CFC" and is followed by a two- or three-digit number. The rightmost digit equals the number of fluorine atoms in the compound; the second rightmost digit equals the number of hydrogen atoms plus one; and the third rightmost digit equals the number of carbon atoms minus one (if zero, it is omitted). Chlorine atoms are assumed to occupy any remaining positions on the molecule. Thus, CFC-12 contains two fluorine atoms, zero hydrogen atoms, one carbon atom, and two chlorine atoms, while CFC-11 contains one fluorine atom, zero hydrogen atoms, one carbon atom, and three chlorine atoms.

[3] To bring attention to the elements constituting a particular compound, the prefix "CFC" is often altered to HCFC (indicating the presence of hydrogen), HFC (hydrogen, no chlorine), and other combinations, as appropriate.

option in new cars in the 1940s, the fraction of new cars equipped with air conditioners grew to approximately 25 percent in 1965, 50 percent in 1970, and 70 percent in 1980. They are now a standard feature in approximately 90 percent of new American cars and 70 percent of light-duty trucks. Combining this growth with the increase in the number of vehicles and the fact that mobile air conditioners are more prone to leak than stationary ones, the annual consumption of CFC-12 for mobile air conditioners reached 50 million kg by 1985, more than one-third of the U.S. consumption in that year.

Foam Plastics

Until World War II, CFCs were primarily used for refrigeration. Then the foam plastics industry got started during the war, when Dow Chemical Company first used CFC-12 as a blowing agent to produce a rigid foam called Styrofoam. CFCs are used to make rigid or closed-cell foams by adding liquid CFC to a polymerizing mixture and then adding heat, which causes the CFC to vaporize and the polymerization reaction to occur. Bubbles created as the CFC evaporates are trapped in the polymer and help to give the product its lightweight but rigid properties. In addition, since CFCs are poor thermal conductors, the CFC bubbles make the material an excellent insulator. Similar principles are used to make flexible, open-cell foams, the primary differences being that flexible foams tend to use CFCs in conjunction with other blowing agents (for example, carbon dioxide) and that the open cells allow the blowing agents to escape within days after manufacture. The cost of the CFC blowing agents represents approximately 20 percent of the manufacturing cost of foams.

Products made with rigid CFC-blown foams range from food containers and beverage coolers to packing materials and building insulation. The demand for CFC-12 as a blowing agent for rigid foams accounted for a near doubling of the amount produced in the late 1940s. In 1959, rigid polyurethane foams blown with CFC-11 were introduced for use as insulation material. Per unit of thickness, these foams provide twice the insulating capacity of fiberglass insulation, and they became the primary insulation material in domestic and commercial refrigerators. Perhaps surprisingly, most home refrigerators typically contain five times more CFCs in their walls (approximately 1 kg of CFC-11) than they use as a refrigerant (approximately 0.2 kg of

CFC-12). In 1985, rigid foam insulation blown with CFCs was used in the construction of approximately 50 percent of new single-family homes and 70 percent of new commercial buildings.

By the 1950s, flexible polyurethane foams were being manufactured using CFC-11 as an auxiliary blowing agent. These flexible foams were widely adopted as cushioning material in products ranging from furniture to automobile seats and dashboards, and their popularity helped to increase sales of CFC-11.

Aerosols

The industry most popularly associated with CFCs is the aerosol industry, which used CFCs 11 and 12, in roughly equal quantities, as propellants. Propellants are liquids stored under pressure in a can that vaporize upon exiting through a nozzle. If the can contains another substance that does not volatilize (such as liquid medication, hairspray, or insecticide), then as the propellant vaporizes it creates a fine spray of the other substance that is known as an aerosol. The relative cost of the propellant depends on the cost of the substance it is dispersing.

During World War II, CFCs were first used as propellants to disperse malaria-fighting pesticides in the Pacific theater. Following the war, CFC-based aerosols were introduced for personal-care products, insecticides, cleaning solutions, asthma medications, and some food products.

Growth of the aerosol industry between the 1950s and 1970s accounted for large increases in CFC production. By 1973, the U.S. aerosol industry consumed more than 200 million kg of CFCs annually, far more than any other industry; the typical American household was estimated to contain forty to fifty aerosol cans, about half of which used CFC propellants (Dotto and Schiff, 1978:146). When concerns about ozone depletion were raised in the mid-1970s, the use of CFCs in aerosols dropped precipitously in the United States and was limited to "essential" uses in 1979. Use in Europe also fell, to about half the level of the late-1970s peak. Consequently, the aerosol industries accounted for only about 10 million kg of CFC production in 1985.

Fire Extinguishants

Also developed during World War II, halons exhibit unique fire suppression capabilities. Halons are like CFCs but contain bromine in place of some of the fluorine or chlorine present in CFCs.[4]

Halons are ideal fire-fighting agents because they quickly suppress fires without conducting electricity, leaving a damaging residue, or harming people. Like CFCs, halons are safely stored under pressure as liquids. In the case of fire, the stored halons are volatilized and can extinguish a fire at low atmospheric concentrations, in part by scavenging free radicals involved in propagating flames. In the 1980s, the cost of the halon in a fire-suppression system typically represented about 20 percent of the total cost of the system. In many cases, halon costs were negligible in comparison with the value of the equipment being protected.

The U.S. Army Corps of Engineers developed halons to suppress fires in tanks and armored vehicles. Two halons have historically dominated the market: halon 1301 and halon 1211. The use of halon 1301 in total-flooding systems began with the military but extended into many civilian uses, such as in computer rooms, museums, vaults, petroleum drilling platforms and pumping stations, and telephone exchanges. In these systems, a precalibrated quantity of halon sufficient to raise its atmospheric concentration to a fire-suppressing level is designed to release into an enclosed space.

Halon 1211 has been widely used in portable fire extinguishers since the early 1970s. Between 1973 and 1984, the demand for halons increased by a factor of four. Nonetheless, the production of halons was substantially smaller than production of CFCs. U.S. consumption of these two halons was approximately 10 million kg in 1987, representing approximately 60 percent of worldwide production.

Solvents

The applications accounting for the most recent growth in CFC production are those that use CFCs as solvents. Solvents are typically used to remove a contaminant from something by transferring the contaminant into the solvent. CFC-113 (trichlorotrifluoromethane) played a major role in the development of the computer industry and a minor role in the dry cleaning industry. Methyl chloroform (1,1,1-trichloroethane) was widely used for degreasing, metal cleaning, and

[4] Halons use a modified version of the CFC naming system. After "halon," the digits from left to right correspond to the number of carbon atoms, the number of fluorine atoms, the number of chlorine atoms, the number of bromine atoms, and the number of iodine atoms. Any rightmost zeros are eliminated, and it is assumed that any remaining positions on the molecule are filled with hydrogen atoms. None of the commercial halons incorporate iodine.

other industrial applications and also, to a limited extent, for dry cleaning. In many cases, most of the solvent could be reclaimed and recycled, and the cost of the solvent typically accounted for a relatively small portion of the product or service cost (less than 5 percent).

CFC-113 was initially used in the aerospace and defense industries to clean critical electronics. Although more expensive than other solvents, its outstanding cleaning ability and mild nature made it the solvent of choice. In 1964, Du Pont introduced CFC-113 as a dry cleaning solvent under the trade name Valclene. Since other solvents for dry cleaning were well established and were relatively less expensive, CFC-113 was not widely adopted by the U.S. dry cleaning industry, but significant quantities were used in Europe for cleaning fine leather goods and furs. Starting in the mid 1970s, the fast-growing computer industry began to use CFC-113 to clean printed circuit boards and high-performance parts made of plastic. Between the mid-1970s and the late 1980s, the U.S. demand for CFC-113 increased by a factor of three and the United States consumed 40 percent of the total 160 million kg produced worldwide. Starting in the late 1970s, some industries using solvents that were more toxic than CFCs or that contributed to the production of photochemical smog began to look more favorably toward substituting CFCs (for example, as metal degreasers).

Methyl chloroform is a substantially less potent ozone depleter than the CFCs, and it was used in large quantities — about 270,000 million kg per year in the United States and perhaps twice that globally — in the late 1970s and mid-1980s. Methyl chloroform is a somewhat more aggressive solvent than CFC-113 and is not compatible with some materials. It has a lower allowable exposure limit than CFC-113 but a higher limit than other chlorinated solvents. Major applications included vapor degreasing and cold cleaning of electronics and other parts, with additional applications in adhesives, aerosols, and coatings.

In addition to these major solvents, CFC-12 has been used as a solvent and carrier for ethylene oxide, which is used to sterilize medical instruments and pharmaceutical supplies.

Production Trends

To meet the growing demands of the industries described above, CFC producers increased their production capacities many times. Worldwide CFC production grew from 500,000 kg in 1931 to more than

1 billion kg in 1987. Figure 3.1 shows the annual worldwide production (excluding communist countries) of CFC-11, CFC-12, and CFC-113, which accounts for approximately 98 percent of total CFC production (AFEAS, 1996). For comparison, worldwide production of halons was on the order of 20 million kg in the late 1980s. CFC production grew rapidly until the mid-1970s, when the possibility that the compounds could deplete stratospheric ozone was first raised and aerosol use plummeted. Production remained stagnant through the recession of the early 1980s, after which it began to rebound until subsequently curtailed by international and domestic regulation.

After Du Pont's patent on CFCs expired in the 1940s, five other U.S. companies began to produce CFCs. Du Pont has been the market leader since it began producing CFCs, and it accounted for approximately 50 percent of U.S. production in 1986. In 1986, Allied-Signal, Pennwalt, Kaiser Aluminum and Chemical, and Essex Chemical (Racon) accounted for approximately 25 percent, 13 percent, 9 percent, and 4 percent, respectively, of U.S. production (Cogan, 1988). An additional producer, Union Carbide, had ceased production in 1977. Another fifteen companies accounted for all of the CFC production outside the Soviet Union and its satellite states: four in Japan, two each

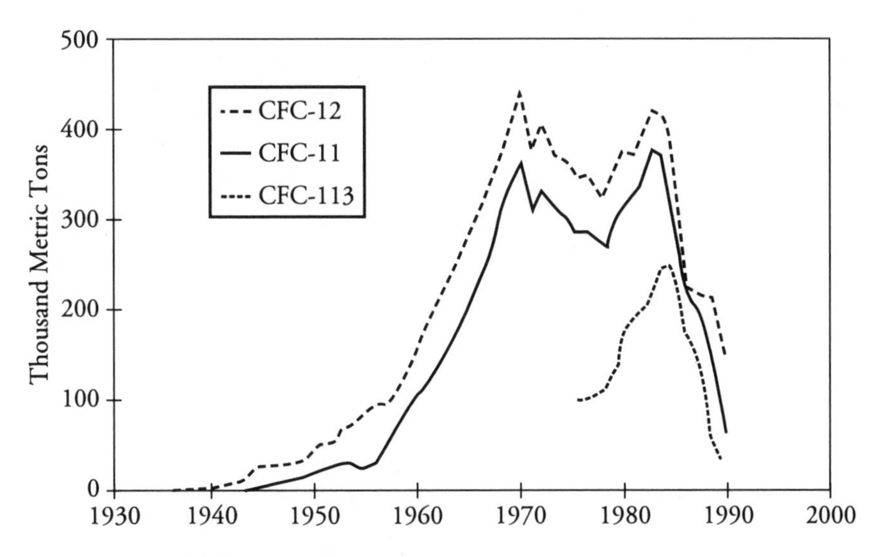

FIGURE 3.1 | Reported global production of CFC-11, CFC-12, and CFC-113 (excludes communist countries).
Source: Data from AFEAS, 1996.

in the United Kingdom and West Germany, and one each in France, Italy, the Netherlands, Canada, Australia, Greece, and India (CMA, 1986). The production of CFC-11 by these other countries surpassed U.S. production in the late 1960s, while U.S. production of CFC-12 was not surpassed until the mid 1970s.

The same industries that use CFCs in the United States also developed in other countries, but usage patterns in other countries differed from those of the United States, as shown in Figure 3.2. In 1985, the United States used a lower proportion of CFCs for aerosols,

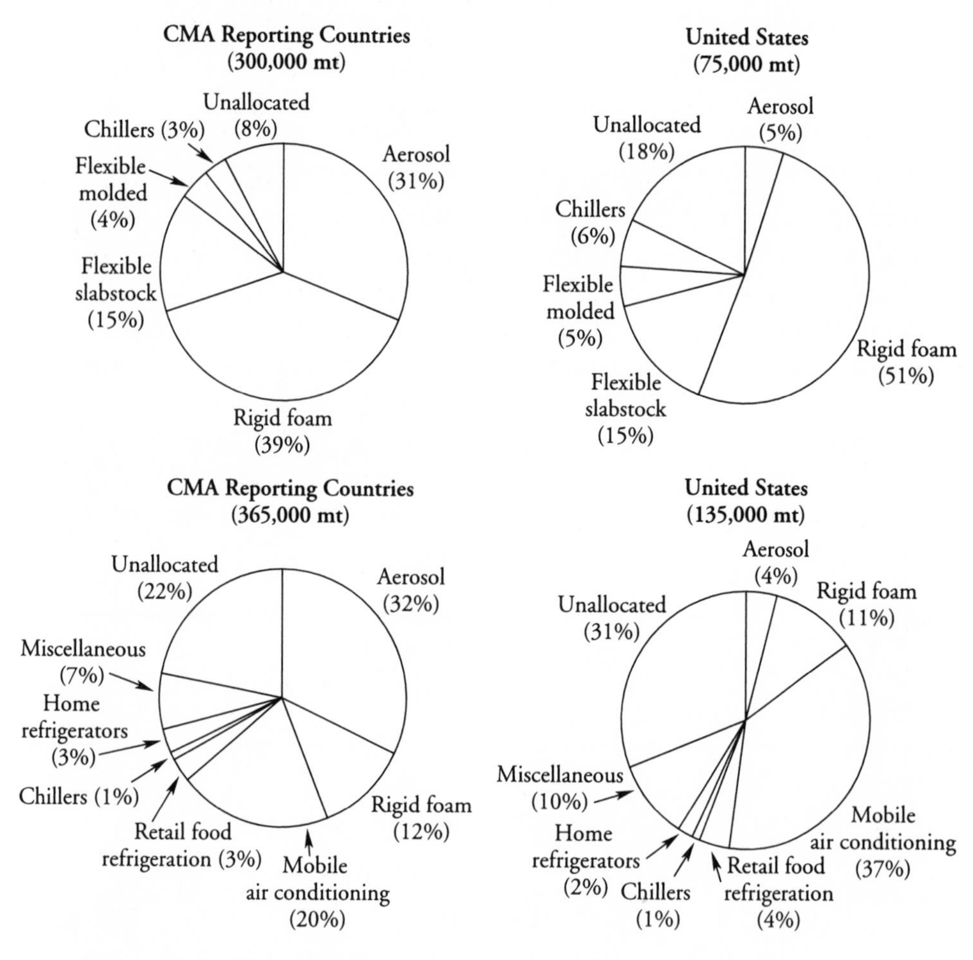

FIGURE 3.2 | Estimated reporting country and U.S. consumption of CFC-11 (top) and CFC-12 (bottom).
Source: Hammitt et al., 1986, pp. 5–6.

and higher proportions for mobile air conditioning and rigid foams, in part because of the aerosol ban and the greater popularity of automobile air conditioning.

Risks and Regulations

CFCs are believed to have two environmental effects that are global in scope: depletion of stratospheric ozone and climate change through enhancement of natural "greenhouse" warming. The effect of CFCs on stratospheric ozone was identified before their effect on climate, and it has been the dominant concern leading to regulation on their use.[5] In addition to its later identification, the climatic effect of CFCs may have been less important to regulation because CFCs are a small contributor relative to carbon dioxide released primarily through fossil-fuel combustion, and because, unlike climate change, all the consequences of ozone loss appear to be adverse.

Initial Concern (1970–1978)

The first concerns about the depletion of stratospheric ozone originated in the 1960s and early 1970s, and did not involve CFCs or halons. At the time, scientists raised the possibility that introduction of water vapor, nitrogen oxides, and other compounds from supersonic transport flights and from solid-fuel launches of the NASA space shuttle could lead to the destruction of stratospheric ozone. Although there was controversy as to which of the chemicals (particularly nitric oxide or water vapor) would be the most destructive, scientists clearly recognized the potential for ozone depletion and began to study the complex chemistry of the stratosphere (NRC, 1975).

Crutzen (1970) proposed a theory that explained how a single molecule of nitric oxide (NO) released into the stratosphere could destroy many molecules of ozone (O_3) in a self-perpetuating cycle. The nitric oxide would first react with ozone, creating nitrogen dioxide (NO_2) and molecular oxygen (O_2). Then the NO_2 could react with free oxygen radicals (O_1) to produce another molecule of oxygen and a molecule of NO, which would start the process over again until some

[5] There are many good histories of the science of ozone depletion and the regulation of chlorofluorocarbons. See, for example, Brodeur (1975, 1986), Dotto and Schiff (1978), Stoel, Miller, and Milroy (1980), Cogan (1988), Roan (1989), Fisher (1990), Benedick (1991), Cagin and Dray (1993), Parson (1993), and Litfin (1994).

reaction occurred that interrupted the cycle. The net effect of this type of reaction would be the conversion of ozone into molecular oxygen.

Although both ozone and oxygen consist only of oxygen atoms, ozone's three atoms provide additional electronic transitions that allow it to absorb ultraviolet (UV) light, an ability that O_2 lacks. Thus it is the ozone in the stratosphere that absorbs UV radiation, and the concentration of stratospheric ozone determines what fraction of incoming UV radiation reaches the earth's surface. Since ozone molecules are continuously produced in the stratosphere as solar radiation dissociates molecular oxygen (O_2), yielding oxygen atoms that subsequently combine with O_2, the atmospheric abundance of ozone is determined by the efficiency of the destruction reactions and the amount of reactive material available. For example, if the destruction reaction is inefficient, there would be little destruction and high ozone levels. Similarly, if the amount of reactive material available (NO in this example) is very small, then again there may be little destruction and no reason for concern.

At the time Crutzen proposed his theory, the destruction of ozone by NO appeared to be highly efficient and the amounts of NO that would be released by the anticipated large global fleet of supersonic transport vehicles appeared to be substantial. Independently, Johnston (1971) estimated that 500 supersonic transports flying seven hours a day for ten years might lead to a 22 percent depletion in the amount of stratospheric ozone and a proportionate increase in harmful UV radiation reaching the earth's surface. Due to the program's high cost and possible environmental problems, Congress voted in 1971 to cease subsidies for the development of American supersonic transport planes. Although similar development efforts continued in the Soviet Union and in Europe, where supersonic jets like the Concorde were developed, the global fleet of these jets has remained relatively small (although more recent interest in supersonic transport has again raised concerns about aerospace-related ozone destruction).

The early controversies and uncertainty about the effects of aircraft and spacecraft on the stratosphere led to investment in more research. Stolarski and Cicerone (1974) concluded that hydrogen chloride (HCl) emitted by the space shuttle's booster rockets would break down in the stratosphere and release chlorine atoms (Cl) that would react with ozone in a self-perpetuating cycle similar to the cycle

Crutzen had proposed for NO. In the case of chlorine, the cycle begins when chlorine reacts with ozone to create O_2 and chlorine monoxide (ClO). The ClO then reacts with O_1 to create more O_2 and chlorine, which reacts again with ozone. Stolarski and Cicerone suggested that ClO could also react with NO to produce Cl and NO_2; this coupling of destructive reactions added uncertainty to the predictions of their theory. They also suggested that because the Cl reaction was six times more efficient than the NO reaction, each chlorine atom reaching the stratosphere would consume an average of 100,000 ozone molecules before being "inactivated" by reacting into some reservoir compound like HCl. These results did not lead to a great deal of concern at the time, because scientists were focusing on the relatively small quantities of chlorine atoms that would be released in the stratosphere by spacecraft and volcanoes, and other means of transporting chlorine to the stratosphere were as yet unrecognized.

Initial discussions about the environmental fate of CFCs occurred in 1972 when industry scientists explored the issue at an Ecology of Fluorocarbons seminar (Cogan, 1988). During the seminar, they identified no cause for concern about CFCs but established a cooperative research program under the Manufacturing Chemists Association (later the Chemical Manufacturers Association). Prior to this seminar, an independent British researcher, James Lovelock, completed a research voyage between Britain and Antarctica. Lovelock had set out to measure levels of CFC-11 in air and seawater to test his theory that CFC-11 might serve as a good tracer of mass transfer processes in the atmosphere and the oceans. His findings (Lovelock, Maggs, and Wade, 1973) implied that CFC-11 behaved essentially as an inert gas and that virtually the entire cumulative production of CFC-11 had distributed globally in the atmosphere. Although he established that CFC-11 had a very long lifetime in the atmosphere, he concluded that the presence of CFCs in the environment appeared to constitute "no conceivable hazard."

In late 1973 Molina and Rowland identified a pathway for the destruction of stratospheric ozone by chlorofluorocarbons (Dotto and Schiff, 1978) and pointed to the importance of the large quantities of CFCs accumulating in the stratosphere. They concluded that the inert and insoluble nature of CFCs meant that they would be unaffected by the natural sinks (oceans, rainwater, particles) that trap other chemicals and facilitate their breakdown in the lower atmosphere. Instead, they said, CFCs would simply diffuse in the atmosphere and be transported to the stratosphere with the global circulation. Only in the stratosphere, when they were hit with ultraviolet radiation, would they decompose, and in decomposing they would release chlorine into the stratosphere that would convert ozone into oxygen. Considering the historical global production of CFCs, millions of tons of CFCs had already been released and were on their way to the stratosphere. But more troubling to Molina and Rowland was the fact that the growth in production of CFCs appeared to be increasing rapidly, doubling every five to seven years (see Figure 3.1). They predicted that if current emission levels continued, the earth's ozone layer would be depleted by 7 percent to 13 percent before it reached a steady state at the end of the next century (Molina and Rowland, 1974).

Cicerone, Stolarski, and Walters (1974) also estimated the time course of stratospheric ozone depletion based on trends in the use of chlorofluorocarbons. They predicted chlorine-catalyzed ozone destruction could exceed all natural sinks of stratospheric ozone by the mid-1980s, and would continue to deplete ozone long after releases of CFCs abated, because of the century-long atmospheric lifetime of the CFCs.

Molina and Rowland presented their concerns at the 1974 meeting of the American Chemical Society. They suggested that the increased exposure to UV radiation, particularly the high-energy UV-B component, could increase rates of human skin cancer, damage plants and reduce yields of food crops, and cool the earth's stratosphere, which might change the climate unpredictably. They called for a complete ban on the use of CFCs as propellants in aerosols, which accounted for over 50 percent of the U.S. consumption of CFCs at that time.

Molina and Rowland received widespread press attention after the conference (Dotto and Schiff, 1978: 21–24). Environmental groups (for example, the Natural Resources Defense Council) began to publicize the threat that aerosol sprays might pose to the atmosphere. The absolute dependence of terrestrial life on the UV-B protection provided by the stratospheric "ozone shield" was emphasized. Congress scheduled hearings to consider regulating or banning CFCs, and the National Academy of Sciences (NAS) initiated efforts to investigate and quantify the threats posed by ozone depletion.

Although Congress did not regulate or ban the use of CFCs

following the hearings in 1974, in 1975 it did appear that some federal action would be forthcoming. Two states acted more quickly. Oregon enacted legislation in the spring of 1975 that banned the sale of spray cans containing CFCs effective in March 1977. In summer 1975, New York passed a requirement that CFC spray cans be labeled to warn consumers of the possible environmental harm caused by CFCs. In general, consumer concern about the environment led to lower consumption of aerosol products and inspired aerosol manufacturers to seek other means for dispensing their products. Indicative of the widespread attention to the issue, Congress received more letters on this topic than on any since the Vietnam War (Dotto and Schiff, 1978: 145).

While consumer concern about aerosol products and an economic recession led to drops in CFC demand, the scientific understanding of the threat posed by CFCs became more complicated. In November 1975 McElroy suggested that the buildup of nitrogen compounds in the atmosphere from the growth in the use of fertilizers could pose a more imminent threat to the ozone layer, depleting it by 10 percent by the year 2015. But in February 1976 Rowland and Molina found that competing reactions between chlorine and nitrogen could interfere with one another in a way that dampened the effects that might be caused by either substance alone (Cogan, 1988).

When the NAS released its report in September 1976, it gave a best estimate of ozone depletion of 7 percent (the low end of the original Molina and Rowland estimate) but indicated that the actual depletion could range between 2 percent and 20 percent. Based on the assumption that most melanoma deaths are due to exposure to solar ultraviolet radiation, the 1976 NAS report estimated that with a 7 percent ozone depletion there might be up to a 15 percent increase in melanoma deaths. Based on a 1975 National Research Council report that had estimated the health effects associated with stratospheric flights, these estimates implied that there would be on the order of a few hundred deaths per year once the ozone depletion had occurred sometime in the next century. Although the 1976 report discussed the evidence suggesting that nonmelanoma (basal and squamous-cell) skin cancers are also related to exposure to UV radiation, the committee did not provide quantitative estimates for the incidence of these cancers. The report also discussed agricultural effects (crop damage), effects on natural terrestrial and aquatic ecosystems (such as harm to phytoplankton), and possible effects of climatic change on the biosphere, and emphasized that any nonhuman biological effects could have significant indirect impacts on humans.

Throughout the 1976 report, a great deal of discussion focused on the uncertainties complicating assessment, and particularly on uncertainty about the magnitude of depletion that would occur. The report concluded that

> The selective regulation of [CFC] uses and releases is almost certain to be necessary at some time and to some degree. Neither the needed timing nor the needed degree can be reasonably specified today because of remaining uncertainties. However, measurement programs now under way promise to reduce these uncertainties quite considerably in the near future. The prospect for narrowing uncertainty and the finding that the rate of ozone reduction is relatively small engender in the committee the conclusion that a one or two-year delay in actual implementation of a ban or regulation would not be unreasonable (NAS, 1976: iii).

Amid this scientific uncertainty, congressional support for regulating CFCs waned. In amending the Clean Air Act in 1977, Congress voted to pass the responsibility for acting to the Environmental Protection Agency. Section 157(b) of the act required the EPA administrator to issue "regulations for the control of any substance, practice, process or activity which in his judgment may be reasonably anticipated to affect the stratosphere, if such effect . . . may reasonably be anticipated to endanger public health or welfare."

In the spring of 1977, the government announced a two-phase plan to limit the use of CFCs. Phase one banned all "nonessential" uses of CFCs in aerosol products after December 15, 1978, and phase two aimed at reducing nonaerosol uses of CFCs. While phase one essentially completed the phasing out process that consumers had initiated in the United States, only three other nations (Canada, Sweden, and Norway) similarly banned the use of CFCs in aerosols. Thus, much of the rest of the world continued to consume relatively high levels of CFCs for use in aerosols. Phase two of the government's plan was not seriously implemented, partly because of the apparent lack of international concern, partly because the public and most policymakers perceived that they had taken care of the problem by eliminating aerosols, and partly because the search for CFC substitutes in other applications had not yielded attractive replacements.

Watchful Waiting (1978–1985)

Between about 1978 and 1985, scientific research continued to yield varying estimates of ozone depletion, as summarized in Figure 3.3 (WMO, 1985). In a report issued in 1979, the NAS increased its estimate of eventual ozone depletion to 16 percent. This increase reflected new evidence that suggested that nitrogen oxides would convert to an inactive form more rapidly than expected and consequently impede the effects of chlorine less than Molina and Rowland had suggested in 1976. Continuing the theme of the 1976 NAS report, the 1979 report also warned of the sharp increases in human skin cancer rates and serious damages to plant and marine life that could occur if the buildup of CFCs continued.

In the early 1980s, international concern motivated many nations to discuss freezing their CFC production capacities and reducing consumption by existing uses. Some nations, notably France and the

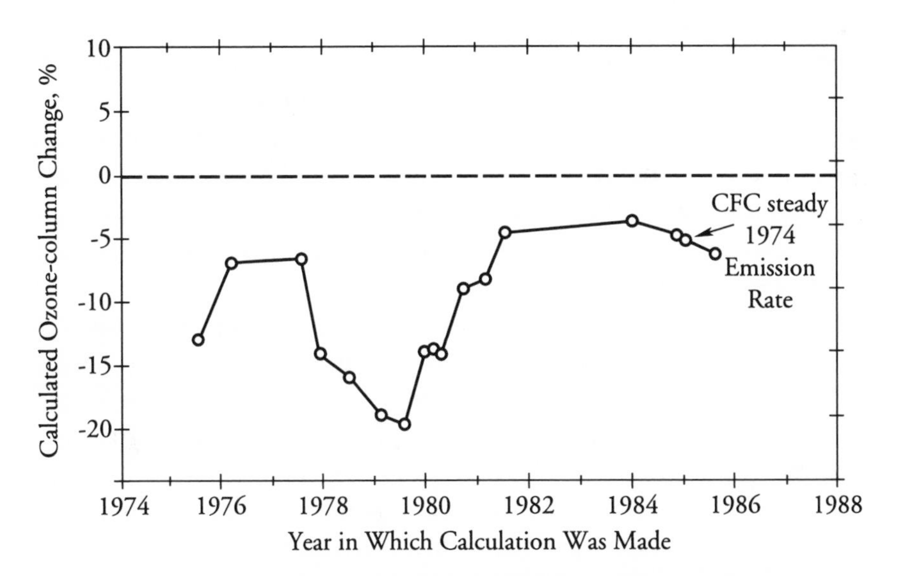

FIGURE 3.3 | Equilibrium column ozone calculated by the then-current version of the Lawrence Livermore National Laboratory model for a scenario with CFC 11 and 12 emissions constant at 1974 rates.
Source: WMO, 1985, p. 773.

United Kingdom, continued to assert that the theories about ozone depletion were still too uncertain to warrant restrictions on CFCs. Nevertheless, in March 1980 the European Community Council of Ministers voted to freeze CFC production capacity and reduce its consumption in aerosols by 30 percent from 1976 levels (Litfin, 1994: 70; Maxwell and Weiner, 1993: 23). In October 1980 the EPA published an advance notice of proposed rulemaking to control domestic CFC emissions (EPA, 1980). But the antiregulatory promise of the newly elected Reagan administration, combined with strong industry reaction against further unilateral regulation, smaller predicted ozone depletion, and the shifting of public attention to the problem of hazardous waste (as at Love Canal) and other environmental concerns led the EPA to delay further rule making on CFCs.

Throughout the early 1980s, scientists continued research on the ability of CFCs to deplete ozone. Much of the research revolved around trying to demonstrate damage to the ozone layer, measuring the lifetimes of different substances, and comparing the abilities of different compounds to deplete ozone. Wuebbles (1981, 1983) developed the ozone depletion potential (ODP) for comparing the abilities of chemicals to deplete stratospheric ozone. The ODP measures the cumulative ozone depletion induced by emission of a kilogram of a specified compound, relative to that induced by emission of a kilogram of CFC-11. The ODP provided a means for comparing regulations on alternative ozone depleters. ODP values suggested that halons could become significant contributors to ozone depletion, despite their relatively low production levels, because of their high ozone-depletion efficiencies (Table 3.1).

In March 1982 the NAS released its third report on ozone depletion. In this report, the NAS estimated global ozone depletion would eventually reach 5 percent to 9 percent, approximately a factor of two below its 1979 estimate. This drop gave the impression that the problem was less serious than had been previously thought. In addition, the 1982 NAS report indicated that the etiology of skin cancer was more complex than previously believed and that consequently the academy was unwilling to make a quantitative estimate of the incidence of melanoma that might result from reduced stratospheric ozone concentrations. The 1982 report added concerns about possible damage to human vision and immune systems. At the time the report was re-

TABLE 3.1 | Atmospheric lifetime, ozone-depletion potential, and global-warming potential for CFCs and related compounds

COMPOUND	ATMOSPHERIC LIFETIME (YR)[a]	MODEL-DERIVED OZONE-DEPLETION POTENTIAL (ODP)[a]	HALOCARBON GLOBAL WARMING POTENTIAL (GWP)[b]
CFC-11	50.0	1.00	1.00
CFC-12	102.0	0.82	3.00
CFC-113	85.0	0.90	1.64
Methyl chloroform	5.4	0.12	35.00
Halon 1211	20.0	5.10	na
Halon 1301	65.0	12.00	1.57
HCFC-22	13.3	0.04	0.37
HCFC-123	1.4	0.014	0.02
HCFC-141b	9.4	0.10	0.14
HFC-134a	14.0	$<1.5 \times 10^{-5}$	0.30

a. Data from WMO (1994).
b. Data from IPCC (1994). GWP (500-year horizon) renormalized using CFC-11 = 1.
na = not available.

leased, CFC sales were down and the public was relatively apathetic. In 1983 the EPA formally considered withdrawing its 1980 notice of proposed rule making, although such action was never taken.

In 1984 the NAS released its fourth report on stratospheric ozone depletion. This time the NAS dropped its estimates of global ozone depletion further, to between 2 percent and 4 percent. These lower estimates were based on the fact that methane, carbon dioxide, and nitrous oxide levels in the atmosphere were increasing and that the presence of these substances would temper the amount of ozone depletion (in part by accelerating global warming, which, by trapping more heat in the lower atmosphere, cools the stratosphere and slows ozone-destruction reactions; Wuebbles, Luther, and Penner, 1983). The NAS report focused attention on the potential contribution of CFCs to global warming and suggested that efforts to reduce ozone depletion would have important ramifications on global warming and vice versa.

Increased Concern and International Action (1985–present)

Following release of the 1984 NAS report, the Natural Resources Defense Council (NRDC) sued the EPA in an attempt to compel the agency to issue final regulations for the proposed 1980 rule making.

While the negotiations between the EPA and the NRDC were ongoing and little action was occurring nationally, delegates from twenty-one nations, including the United States, met in Vienna in March 1985. At this meeting, the delegates adopted the Vienna Convention for the Protection of the Ozone Layer, which established a program for cooperative international research and monitoring. The delegates also scheduled a second meeting for spring 1987, when international measures to limit the use of CFCs would be discussed.

In May 1985, two months after the Vienna Convention, a British Antarctic survey team led by Joe Farman reported the first observed damage to the ozone layer, (Farman, Gardiner, and Shanklin, 1985). Measuring ozone above Antarctica, the team first observed the Antarctic "ozone hole" above its research station in 1981 (Somerville, 1996). However, these findings seemed so shocking, given that satellite observations had not revealed the hole, that the scientists waited to reconfirm their observations in subsequent years before reporting them. Shortly after Farman's team announced its results, NASA scientists confirmed them, using the total ozone mapping spectrometer (TOMS) data collected by the Nimbus-7 satellite.[6] The NASA scientists were also able to create pictures showing that the hole extended over the entire Antarctic region; these pictures appeared widely in the popular press and presented the public with striking visual images of the effects of ozone depletion.

While news of the ozone hole revived public concern about ozone depletion and its possible catastrophic consequences, it sent scientists into a frenzy by demonstrating their fundamental lack of understanding of the complex chemistry and physics of the upper atmosphere. None of the models that had been developed to date had predicted so much ozone depletion would be possible so soon, and no one was able to explain the development of the hole over Antarctica. In addition, the hole appeared to be a temporary phenomenon, appearing in September and disappearing by November. Several competing hypotheses, based on chemical and dynamic factors, were advanced.

[6] It is widely reported (for example, in Somerville, 1996) that NASA failed to discover the hole because its data-processing algorithms discarded readings beyond a specified range on the assumption that these would be erroneous. Pukelsheim (1990) reports that NASA had investigated the anomalously low springtime Antarctic results but concluded they were incorrect because they conflicted with much higher readings from the station at the South Pole, which itself was subsequently shown to have been reading too high.

In October 1985, while scientists again were observing the Antarctic ozone hole, a consent decree was issued settling the lawsuit between the NRDC and the EPA. The consent decree required the EPA to decide by 1987 whether or not to regulate nonaerosol uses of CFCs. The EPA announced in January 1986 that it was preparing a stratospheric ozone protection plan.

In the fall of 1986, several teams of scientists assembled in the Antarctic to study the ozone hole and to try to explain its development. The teams were unsuccessful in deciphering the origins of the hole, but they were able to rule out dynamic hypotheses, as these required transport in the lower stratosphere that was inconsistent with observations (personal communication, D. Wuebbles), and they collected large quantities of data for future analysis.

Although the source of the ozone hole remained uncertain, delegates convened in Geneva in April 1987 as scheduled under the Vienna Convention. During the Geneva meetings, the delegates reached an agreement to freeze CFC production at 1986 levels beginning in 1990, to reduce CFC production 20 percent by 1992, and to discuss options for controlling halons and for further reductions of CFCs. A meeting was scheduled for September 1987 in Montreal, at which the delegates planned to sign a protocol agreement. Shortly after the groundwork for the Montreal Protocol had been laid, Reagan's interior secretary, Donald Hodel, and White House science adviser William Graham proposed that people could adopt "personal protection" like hats, sunglasses, and suntan lotion to reduce their exposure to UV radiation, and they suggested that CFC production cuts might not be necessary. The plan was quickly dubbed the "Rayban plan" by environmentalists, who pointed out that it would do little for crops and wildlife — "It's very hard to get fish to wear sunscreen," said David Doniger of the NRDC (Cogan, 1988: 52) — and they estimated the costs of such a plan to be $10 billion a year, an amount that was ten times higher than the projected costs of a freeze in CFC production.

In June 1987 President Reagan formally endorsed the State Department's position supporting government controls on CFCs. Shortly afterward, in September 1987, the Montreal Protocol on Substances that Deplete the Ozone Layer was signed, and delegates agreed to continue meeting periodically to review the adequacy of the protocol. Later in the year, the EPA released a regulatory impact analysis (described below) that estimated that millions of cases of skin cancer among Americans would result over the next century if CFC and halon emissions were not controlled. This probably represents one of the highest population-risk estimates put forth by EPA. The agency concluded that regulations to implement the Montreal Protocol were highly cost-effective. It issued draft regulations in December 1987 and announced plans to issue final regulations by August 1988 (EPA, 1987a). In addition to the EPA analysis, the Council of Economic Advisers prepared an analysis that found that the benefits of additional CFC regulations vastly exceeded their costs. This analysis was apparently influential within the executive branch, although it was not circulated externally (Benedick, 1991: 63; personal communication, S. DeCanio).

The terms of the Montreal Protocol differed from the preliminary Geneva agreement, requiring slower but deeper cuts in CFC consumption. The protocol provided that nations freeze CFC consumption at 1986 levels in 1989, cut them 20 percent beginning in 1993, and cut them an additional 30 percent by 1998 (developing countries are allowed a ten-year grace period before meeting similar conditions). The protocol also froze annual consumption of halons at the 1986 level beginning in February 1992 (the Montreal Protocol is summarized in Benedick, 1991, and reprinted in EPA, 1987a).

In fall 1987, preliminary results from the data collected during the 1986 Antarctic expedition suggested that there was a connection between CFCs and the appearance of the ozone hole. At the same time, scientists returned to the Antarctic in a second attempt to unravel the mysteries of the hole. The 1987 ozone hole was both larger and deeper than had been previously observed. Scientists postulated that surface reactions on the polar stratospheric clouds present during the Southern Hemisphere's winter months might explain its development (Solomon et al., 1986). Polar stratospheric clouds develop when the air mass over the Antarctic rotates in the "polar vortex," effectively isolating polar air from the rest of the atmosphere. Extremely low temperatures in the vortex cause water vapor and other chemicals to freeze into ice crystals. Laboratory work by Molina and coworkers subsequently demonstrated that the reaction between two important sinks for chlorine, chlorine nitrate and hydrogen chloride, was greatly increased in the presence of ice crystals like those present in the polar stratospheric clouds (Molina et al., 1987). The products of this reaction are two chlorine atoms that react with ozone, and nitric acid, which prohibits the inhibiting reaction between nitrogen dioxide and chlorine atoms. In short, the polar stratospheric clouds create some of the best possible conditions for ozone depletion, and lower levels of ozone (which mean cooler temperatures) create some of the best possible conditions for polar stratospheric clouds.

In March 1988, NASA published the Executive Summary of the Ozone Trends Panel Report. This report provided the first documented evidence of global ozone loss, and at rates exceeding those predicted by existing atmospheric models. It combined the efforts of more than one hundred scientists and represented a clear indictment of CFCs. The report concluded that between 1969 and 1986, average ozone levels dropped 1.7 percent to 3 percent in the temperate Northern Hemisphere, with wintertime losses averaging 2.3 percent to 6.2 percent. Although these losses were not necessarily due to CFCs, they were consistent with the theory of CFC-caused depletion, albeit somewhat larger than predicted by current models. The report also implicated CFCs as a primary cause of the Antarctic ozone hole and recognized the hole's hemispheric effects, documenting that ozone-depleted air moves toward the equator when the polar vortex breaks up in the Southern Hemispheric spring, causing ozone declines of 5 percent or more in southern latitudes poleward of 60 degrees.

With the observation of widespread ozone depletion, concern shifted to international negotiations and the issue of whether regulations would be strong enough. After its signing in September 1987, the Montreal Protocol required ratification by at least eleven of the signatory nations, representing two-thirds of global CFC consumption, before it would go into effect. In March 1988 the United States became the second nation, following Mexico, to ratify the protocol, which entered into force January 1, 1989. In the United States, the EPA is charged with implementing the requirements of the Montreal Protocol, and it issued regulations in August 1988 to initiate the process. The protocol requires periodic review of scientific and technical issues, with the expectation that its terms will be modified as information about CFCs and their effects on the atmosphere develop. To date, three such revisions to the Montreal Protocol have occurred, the 1990 London, 1992 Copenhagen, and 1995 Vienna amendments. These amendments have accelerated the phaseout of some substances and included restrictions on additional ozone-depleting substances. Table 3.2 summarizes the changes. In March 1994 the EPA issued a list of "acceptable" substitutes for CFCs under its Significant New Alternative Policy (SNAP) program (EPA, 1994).

| EPA's Regulatory Impact Analysis

While the 1987 ozone-hole research was underway, the EPA was completing its regulatory impact analysis and preparing to decide whether and how to regulate CFCs. Published in December 1987, the analysis provided a comprehensive assessment and quantification of the costs and benefits to the United States of national and international regulation of ozone-depleting substances. It concluded that the roughly 50 percent cut in CFC emissions and stabilization of halon emissions proposed by the Montreal Protocol were extremely cost-effective, with U.S. benefits of $6.5 trillion and costs of only $27 billion. However, these estimates are subject to large uncertainties.

Because of the long periods over which CFCs and halons deplete stratospheric ozone, the EPA analysis required assessing the effects of regulation over a much longer period than is conventionally required for other pollutants. This long horizon contributes to uncertainty about the effects of regulations and also highlights methodological issues in assessment, such as the population to be included and the appropriate discounting of future impacts. The EPA adopted some innovative approaches to dealing with these issues. In quantifying health effects, it considered skin cancers and other effects accruing to a reference population consisting of Americans living at the time of the analysis plus those who would be born by 2075, although it apparently counted only cancers that would occur before 2075. In developing monetary estimates of the value of averting mortality from these skin cancers, the EPA accounted for increases in the value of health that may accompany increases in material well-being by inflating the value of a statistical life saved through time, thus partially or totally offsetting the effect of discounting over these long periods.

The EPA analysis considers future CFC and halon emissions,

TABLE 3.2 | Limitations on consumption of ozone-depleting substances: Montreal Protocol and amendments[a]

	MONTREAL PROTOCOL (SEPTEMBER 1987)[b]	LONDON AMENDMENT (JUNE 1990)[c]	COPENHAGEN AMENDMENT (NOVEMBER 1992)[c]	VIENNA AMENDMENT (DECEMBER 1995)
CFCs	*1989*: 100% of 1986 level *1993*: 80% of 1986 level *1998*: 50% of 1986 level	*1995*: 50% of 1986 level *1997*: 15% of 1986 level *2000*: 0	*1994*: 25% of 1986 level *1996*: 0	
Halons	*1992*: 100% of 1986 level	*1995*: 50% of 1986 level *2000*: 0	*1994*: 0	
Carbon tetrachloride		*1995*: 15% of 1989 level *2000*: 0	*1996*: 0	
Methyl chloroform		*1993*: 100% of 1989 level *1995*: 70% of 1989 level *2000*: 30% of 1989 level *2005*: 0	*1994*: 50% of 1989 level *1996*: 0	
HCFCs		Identified as transitional substances to be phased out by 2040 or, if possible, by 2020	*1996*: 100% of 1989 HCFCs + 3.1% of 1989 ODP-weighted CFCs *2004*: 65% of 1996 cap *2010*: 35% of 1996 cap *2015*: 10% of 1996 cap *2020*: 5% of 1996 cap *2030*: 0	
Methyl bromide			*1995*: 100% of 1991 level	*2001*: 75% of 1991 level *2005*: 50% of 1991 level *2010*: 0

a. Consumption defined as production plus imports minus exports. Developing countries are allowed to delay restrictions beyond deadlines shown.

b. Restrictions apply to years beginning July 1.

c. Restrictions apply to years beginning January 1.

ozone depletion, consequences with and without alternative sets of regulations, and the economic costs of the alternative regulations. It begins with an estimate of CFC and halon emissions through 2075 in the absence of the Montreal Protocol or other new regulations. Based on several analyses of historical trends and the broad economic determinants of CFC and halon use, "low," "medium," and "high" emission scenarios were considered. As shown in Table 3.3, emissions of CFCs 11, 12, and 113 and halon 1301 were projected to grow between about 2 percent and 6 percent per year for the period from 1985 to 2000, and between 1.3 percent and 4.7 percent per year for the period from 2000 to 2050. Halon 1211 was projected to grow at a somewhat greater rate; the growth in emissions of all compounds was conservatively projected to end by 2050.

Time paths of total global ozone depletion corresponding to these scenarios were calculated using a simplified representation (Connell,

TABLE 3.3 | CFC and halon emission scenarios and calculated ozone depletion

	LOW GROWTH	MEDIUM GROWTH	HIGH GROWTH
PROJECTED EMISSIONS GROWTH RATE (%/YR)			
1985–2000			
CFC-11, 12	2.1	3.6	5.1
CFC-113	2.1	4.0	6.1
Halon 1301	−0.5	1.1	2.0
Halon 1211	5.5	8.8	12.0
2000–2050			
CFC-11, 12	1.3	2.5	3.8
CFC-113	1.3	2.5	3.8
Halon 1301	1.6	3.2	4.7
Halon 1211	1.5	2.9	4.4
2050–2075			
All gases	0.0	0.0	0.0
PROJECTED OZONE DEPLETION (%)			
1990	0.27	0.27	0.28
2010	1.25	1.71	2.33
2030	2.55	4.86	9.13
2050	4.32	12.3	>50.0
2075	7.09	39.9	>50.0

Source: EPA, 1987a.

1986) of contemporary atmospheric models that represent a globally averaged "one-dimensional" atmosphere — that is, the models represent an idealized column of the atmosphere, with insolation and other input conditions representing some average of the latitudinally and temporally varying conditions in the real atmosphere. Although two-dimensional models (representing the effects of latitude and altitude) where available at the time, the EPA judged these to be insufficiently reliable for its analysis. Even holding the atmospheric model constant, the range of future emissions was so large that projected ozone depletion ranged from about 4 percent to more than 50 percent by 2050 (Table 3.3). As illustrated in Figure 3.4, ozone depletion for the medium emission scenario was projected to increase modestly until about 2030 or so, when nonlinearities in the atmospheric response would lead to a "cliff." In contrast, depletion was projected to be minimal under the 50 percent reduction in CFC emissions and the freeze on growth in halon emissions called for by the Montreal Protocol.

Next, the effects of these alternative scenarios of ozone depletion on a variety of end points, including human health, materials, and

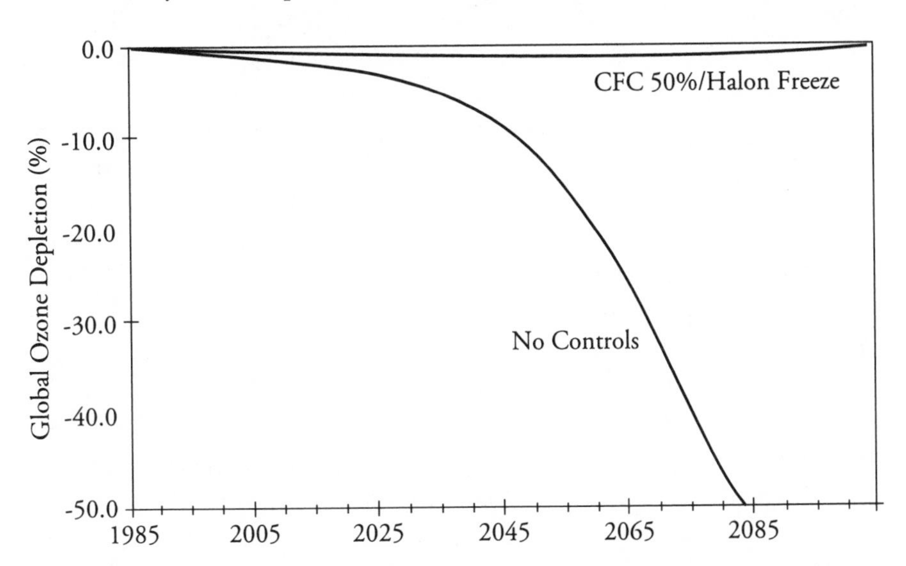

FIGURE 3.4 | Global ozone depletion estimates under the medium emission scenario, for the no-controls case and for the CFC 50% reduction/halon freeze case.
Source: EPA, 1987b, pp. 6–5.

plants and aquatic organisms, were considered. (Because projected ozone depletion under the primary control scenario was close to zero, nearly all the damages due to ozone depletion would be offset by controls.) The estimated benefits are summarized in Table 3.4. Despite the range of end points considered, the estimated monetary benefits are dominated by the effect on the death rate from skin cancer.

The effect of ozone depletion on UV-B radiation reaching the earth's surface varies with frequency, the presence of ground-level ozone that can also absorb UV-B, cloudiness, and other factors. The EPA developed a model to represent the effect of decreased stratospheric ozone on UV-B flux by frequency, then weighted the frequencies in proportion to their biological activity and summed the changes into an effective change in a biological action spectrum.[7] The result was that for ozone depletion of no more than 20 to 30 percent, the increase in biologically effective UV-B flux was about double the ozone depletion, with an increasing proportional response of UV-B to ozone depletion for higher depletion levels (EPA, 1987b: Appendix E).

The evidence linking UV-B flux to nonmelanoma (basal and squamous-cell) skin cancer is fairly strong: these cancers are located primarily on parts of the body that are routinely exposed to sunlight, and their incidence (primarily among Caucasians) is higher among outdoor workers and in populations living nearer the equator. Based on epidemiological evidence, the EPA developed an exposure-response model in which incidence increases roughly in proportion to the increase in UV-B flux. Because nonmelanoma skin cancer is quite common (without ozone depletion, the average risk for white Americans is about 30 percent; EPA, 1987b; Appendix E), projected increases quickly lead to millions of additional cases: the EPA estimated 153 million additional cases to its reference population under the medium ozone-depletion scenario. Fortunately, these cancers are highly treatable, so even with no assumption of improvements in medical technology, the projected increase in mortality was only 3 million.

As discussed by the National Academy of Sciences in its 1979 report, the evidence linking melanoma cancer incidence to UV-B is much weaker than that for nonmelanoma cancers, but the risk appears to increase with acute episodes of sunburn, especially in childhood.

[7] Two such spectra, representing effects on DNA and on erythema, were developed. Their quantitative responses to ozone depletion are similar.

TABLE 3.4 | Summary of estimated benefits of regulations

IMPACT	VALUE (IN BILLIONS)
Skin cancer mortality	$6,350.0
Skin cancer morbidity	61.3
Cataracts	2.6
Crop damage from UV-B	23.4
Crop damage from ground-level ozone	12.4
Reduction in commercial fish harvest	5.5
Damage to plastics	3.1
Sea-level rise (global warming)	4.3
Total	$6,300.0

Note: Total as reported in source.
Source: EPA, 1987a.

The EPA estimated an increased incidence of 782,000 cases and, because melanoma is much more frequently lethal, 187,000 additional deaths for its middle depletion scenario. Combined with the 3 million additional deaths from nonmelanoma cancer, prevention of these skin-cancer deaths accounts for 98 percent of the quantified benefits of regulation. The present value is $6.35 trillion, using a value of averted cancer death of $3 million that grows 1.7 percent annually, and discounting the benefit at 2 percent per year (for an effective discount rate of about 0.3 percent per annum).[8] The value of averting the morbidity associated with increased skin-cancer incidence, $61.3 billion for the middle case, is based on the cost-of-illness method, by which the loss associated with disease is quantified as the value of the resources consumed in treating it (including lost work time). The estimated morbidity costs are about $15,000 per case for melanoma and $5,000 per case for nonmelanoma skin cancer.

Increases in cataracts are quite plausibly linked to increases in UV-B radiation. The EPA estimated 18.2 million additional cataracts in its reference population. Using cost-of-illness and individual willingness-to-pay measures, the value per cataract case was estimated as $15,000, the same as for a case of (nonfatal) melanoma. The possibility that

[8] These figures are for the EPA's middle case. In its low case, the EPA assumes a current value of life of $2 million, growing at 0.85 percent per year, and discounts at 6 percent per year. In its high case, the value of life is $4 million growing at 3.4 percent per year with a discount rate of 1 percent. In the high case, the present value of averting a skin-cancer death is larger for future generations than for the present: it grows about 2.4 percent annually.

increased UV-B would also lead to suppression of the human immune response and so increase the incidence of a range of diseases was recognized but not quantified, due to inadequate data.

In addition to human health effects, the benefits of reducing a number of other UV-related damages were at least partially quantified. These included damage to crops from UV-B and from higher concentrations of ground-level ozone (produced by the interaction of UV radiation, hydrocarbons, and nitrogen oxides), damage to commercial fishing because of increased larval mortality from UV-B exposure, increased weathering of outdoor plastics, and sea-level rise due to global warming. As summarized in Table 3.4, these nonhealth effects did not significantly contribute to the overall benefit estimate.

The costs of compliance with the Montreal Protocol restrictions were estimated using a detailed model of technological substitution and include reductions in both consumer and producer surplus. More than 550 control options were evaluated, including chemical and product substitutes, engineering controls on emissions, recovery and recycling of CFCs and halons, and changed work practices to reduce leakage. Because many of these options were not in use at the time, cost, effectiveness, and even availability were uncertain. To some extent, the EPA incorporated these uncertainties by dividing the options into classes based on their date of assumed availability, and considered delays in availability on the order of five years in sensitivity analyses.

Costs were reported as present values over the period from 1989 to 2075, discounted at 2 percent per year, in constant 1985 dollars. An idealized "least cost" estimate of $27 billion in costs of controlling CFC emissions was based on the assumption that firms would adopt each control measure as soon as it became cost-effective. Alternative analyses assuming additional delays in adopting cost-effective control measures associated with lack of information, imperfect capital markets, managerial inertia, and the like, together with delayed availability of several important technologies, substantially increased estimated control costs. However, even the highest of these estimates, $38 billion, remains far below the estimated benefits. Estimated costs are effectively capped by the assumption that the substitute chemicals HCFC-123 and HFC-134a would efficiently replace CFCs 11 and 12 in foam-blowing and refrigeration applications at a price increase of

$5.48/kg (a substantial increase from the unregulated price of about $1.50/kg); this assumption may not have allowed adequate consideration of specialized niche applications (personal communication, T. Vogelsberg).

Risks Posed by Global Climate Change

As noted above, the 1976 NAS report discussed the role of CFCs in climatic change. CFCs contribute directly to global warming through their role as greenhouse gases. Like carbon dioxide, the primary greenhouse gas, CFCs increase the net radiative forcing or heat-trapping effect of the atmosphere by absorbing infrared heat radiated from the earth and reemiting it to the lower atmosphere. Per kilogram, CFCs are roughly 1,000 to 10,000 times more efficient in trapping infrared heat than is carbon dioxide (IPCC, 1994). The report concluded that "a condition of long-continued growth of [CFC] releases, of as little as a few percent per annum would, to the best of our present knowledge, lead ultimately, perhaps in a century or two, to climatic change of drastic proportions" (NAS, 1976: 65).

The 1984 NAS report discussed the climate change issue in more detail and focused on the complex interactions between ozone depletion and climatic change. As mentioned previously, this report indicated that solutions aimed at reducing global climate change could affect ozone depletion and vice versa, and that efforts to deal with these global problems should be coordinated.

In addition to their direct contribution to radiative forcing, CFCs indirectly reduce radiative forcing by depleting stratospheric ozone, which itself increases greenhouse warming. Initially, this indirect effect was not thought to be as significant as the direct effect, but in the early 1990s it was determined that, at least on a global basis, the indirect cooling might roughly cancel the direct warming (although the effects would not necessarily cancel regionally, because of latitudinal variation in ozone depletion and in radiative forcing). The larger than anticipated indirect effect results because ozone depletion is both greater in magnitude and occurring at lower altitudes (where ozone is a more potent contributor to greenhouse warming) than had been expected in earlier studies (Daniel, Solomon, and Albritton, 1995).

Concern about global warming has led to a number of international meetings, including the 1992 United Nations Conference on

Environment and Development held in Rio de Janiero. The Framework Convention on Climate Change was signed at this meeting, committing signatories to the goal of limiting carbon dioxide and other human-related greenhouse gas emissions to their 1990 levels by 2000 and preparing "national action plans" detailing the steps by which emission reductions will be achieved. CFCs were specifically excluded from the Framework Convention because they were being controlled under the Montreal Protocol and its subsequent amendments.

| Industrial Response

Industrial responses to concern about the use of CFCs began in 1974 and also varied among the three periods. The producer industry has consistently participated in scientific and policy discussions, while user industries have generally played a less active role.

Initial Concern (1970–1978)

The possibility that CFCs might represent an environmental danger did not become a serious issue for industry until 1974. When the theory that CFCs might deplete ozone then surfaced, consumers focused on aerosol products, and the aerosol-using industries quickly began to develop alternative techniques for dispensing their products, such as pump sprays and roll-ons. As early as June 1975, Johnson Wax announced it would no longer use CFCs in aerosol products (Dotto and Schiff, 1978: 165–166). Since other chemicals (such as hydrocarbons) were already available, were often less expensive, and were in fact widely used as propellants, many industries were able to switch propellants relatively easily. The main issues to consider in switching propellants were the differences in cost, product quality, toxicity, flammability, and reactivity of the new propellant when compared with the CFC it was replacing.

CFC producers responded to the initial concerns in three ways. First, the producers vigorously contested the theory of Molina and Rowland and the view that it warranted interference with highly useful products (Dotto and Schiff, 1978). They called for further research, but indicated their willingness to cease production if new research should show that action was necessary. The 1974 statement of Raymond McCarthy, Du Pont's technical director, before a congressional subcommittee clearly summarized Du Pont's position. He stated that "if creditable scientific data developed in this experimental program show that any chlorofluorocarbons cannot be used without a threat to health, Du Pont will stop production of these compounds" (Dotto and Schiff, 1978: 180). The other producers took similar positions, and collectively they held that it would be premature to enact legislation in light of the considerable scientific uncertainty that existed at the time. In support of this position, producers offered substantial funding for research to resolve existing uncertainties. Finally, several producers also initiated research and development programs to look for substitutes.

Since the aerosol industry was the initial focus of concern, other industries did relatively little to reduce their use of CFCs. In general, these industries were looking to producers to identify "drop-in" substitutes, alternative compounds that could be used to replace CFCs in existing equipment. No quick and simple substitutes were identified in the late 1970s, and since the nonaerosol industries were not pressured to reduce CFCs by regulatory action, consumption of CFCs by these industries held relatively steady or grew slightly during this period. The decline in consumption of CFCs for aerosols explains the drop in production that occurred during this period, shown in Figure 3.1 (Hammitt et al., 1986).

Watchful Waiting (1978–1985)

By the early 1980s, the U.S. phaseout of CFCs in "nonessential" aerosols was complete, and aerosol use had also declined in Europe (Dotto and Schiff, 1978: 171). In addition, new scientific findings appeared more optimistic and there was a general deregulatory climate in the United States. These factors, combined with overcapacity due to reduced demand (that resulted from the aerosol phaseout and a general economic downturn) and the fact that none of the identified substitutes could compete with CFCs in cost in the absence of regulation, led companies to shelve research and development efforts for substitutes. Producers continued to support scientific research into the effects of CFCs, but there was little incentive for industries to seek alternatives to CFCs or alternative technologies. As the country came out of recession, production of CFCs returned almost to their pre-1974 levels, and growth in production and consumption of CFC-113 rose dramatically by the end of this period (Figure 3.1).

Increased Concern and International Action (1985–present)

Following the discovery of the ozone hole, U.S. CFC producers shifted their position to support international controls to limit growth in CFC production. In September 1986, a spokesman for the Alliance for Responsible CFC Policy announced: "Responsible policy dictates, given the scientific uncertainties, that the United States work in cooperation with the world community under the auspices of the United Nations Environment Programme (UNEP) to consider a reasonable global limit on the future rate of growth of fully halogenated CFC production capacity" (Cogan, 1988: 50). A few days later, Du Pont sent a letter to its customers indicating its change in position and that it had restarted its work on substitutes. Du Pont wrote that "if the necessary incentives were provided, we believe that [CFC alternatives] could be produced in volume in a time frame of roughly five years" (Cogan, 1988: 50).

While the American CFC producers had changed their position to support limits on growth, they were unified in their conviction that any limits had to be international. This position became clear when a representative of the American CFC industries commented on two U.S. Senate bills aimed at reducing domestic use of CFCs. The industries argued that trying to force international action by threatening unilateral domestic action not only placed an unfair burden on U.S. industry but also had failed when the United States banned most uses of CFCs in aerosols in the 1970s, and that such an approach would fail again. Their position prevailed, and ultimately domestic action has followed international requirements.

At the time, there was suspicion that the U.S. producers had shifted their position to support international controls because they believed they had a competitive advantage over foreign producers in developing and marketing substitute compounds. Indeed, the European Community suspected that the only reason the antiregulatory Reagan administration supported CFC controls was that U.S. producers had secretly developed substitutes (Litfin, 1994: 108).

As the Montreal Protocol approached, regulations to substantially reduce CFC production looked imminent. Nonetheless, most of the user industries continued to count on development of substitutes by producers. In Senate hearings held in May 1987, producers still focused on uncertainty about levels of ozone depletion. At the same hearings, the EPA presented preliminary findings of its risk assessment efforts and suggested that the number of skin cancer deaths would increase if no action were taken. Although no action was taken immediately following the hearings, in February 1988 Senator Robert Stafford (R-Vt.), who had been involved in the hearings and had since become convinced that the threat of CFCs to health was no longer in doubt, sent a letter to the Du Pont chairman suggesting that it was time for Du Pont to make good on its 1974 pledge. Du Pont responded that "at the moment, scientific evidence does not point to the need for dramatic CFC emissions reductions. . . . A precipitous reduction in CFC supplies would be both unnecessary and highly disruptive." However, eleven days after Du Pont responded to Senator Stafford, NASA's Executive Summary of the Ozone Trends Panel report was released. Since the report detected a global decline in ozone levels and identified CFCs as the likely cause, on March 24, 1988, Du Pont made good on its pledge and announced plans to phase out production of CFCs. On the same day, Pennwalt also urged a phaseout of CFCs. These companies and Allied-Signal turned their efforts toward the production of substitutes. Of the two remaining producers, Kaiser decided to quit the business, selling its assets to LaRouche Chemicals, and Essex Chemical (Racon) indicated it would continue to produce to meet existing demand (for servicing refrigeration systems, for example) as long as it could do so (Cogan, 1988).

Substitutes: Toxicity and Cost Tradeoffs

Two broad classes of alternatives to CFCs and halons are currently being pursued: substitute chemicals and not-in-kind alternatives. Unlike the case with the earlier phaseout of CFCs in aerosols, most of the CFC substitutes in other applications are likely to have increased costs, although cost increases appear to have been smaller than forecast at the time of the Montreal Protocol (Cook, 1996; Hammitt, in press).

Substantial research has gone into the development of alternative compounds. In some cases, alternative compounds can be used as drop-in substitutes in existing refrigeration, air-conditioning, cleaning, and other equipment. In others, the alternative compounds are not compatible with existing equipment, so substitution requires retrofitting existing equipment or using the substitute compounds only in new equipment. In refrigeration and air-conditioning applications, for example, alternative seals are often required (because of incompatibility with current lubricants). Not-in-kind substitutes include alternative processes, such as alternative refrigeration cycles, and use of different soldering fluxes and cutting oils in "no-clean" manufacturing processes.

The essential criteria for substitutes are that their behavior must be similar to the compounds they are replacing, and either that they contain no chlorine or bromine (which destroy ozone) or that they break down in the lower atmosphere (which generally means they contain at least one hydrogen atom and so are not "fully halogenated"). Perhaps not surprisingly, most of the new compounds that have been developed, hydrochlorofluorocarbons (HCFCs) and hydrofluorocarbons (HFCs), are chemically similar to CFCs.

HCFCs and HFCs have several major disadvantages as CFC substitutes. First, they are more expensive to produce — typically two to five times the cost of CFCs 11 and 12. This higher cost reflects scale economies, the higher cost of fluorine than chlorine, and the required multireaction process (personal communication, A. McCulloch). In addition, because they contain chlorine, HCFCs maintain some (reduced) potential to deplete stratospheric ozone. For example, HCFC-22 (which has been used in home air conditioners for many years) has an ozone depletion potential of about 0.04, and HCFC-141b, a new compound, has an ODP of about 0.1 (Table 3.1). Because HCFCs deplete stratospheric ozone, they are viewed as "transitional" chemicals to be used until better alternatives are developed. The inclusion of hydrogen makes HCFCs more reactive than CFCs, reducing their atmospheric lifetimes and ODPs by about a factor of ten. In contrast, HFCs do not contain chlorine and consequently do not deplete stratospheric ozone. With respect to climate change, HCFCs, and HFCs are approximately as potent as CFCs in terms of their instantaneous radiative effect, but their atmospheric residence times are often shorter, and consequently their total direct contribution to global climate change is reduced, as measured by global warming potential (Table 3.1).

A concern about the substitute compounds was that they might be more toxic to humans, because they are chemically less stable than the CFCs they replace. Consequently, toxicity testing for the substitutes is an important step in their approval and ultimate consumer acceptance. To deal with the costs of toxicity testing, thirteen CFC producers from the United States, Europe, and Japan formed the Program for Alternative Fluorocarbon Toxicity Testing (PAFT). The EPA's Significant New Alternative Program (SNAP) was established to approve substitutes for particular applications; it published an initial list of approved and disapproved substitutes in March 1994 (EPA, 1994). Of the new compounds, only HCFC-123 has a chronic exposure value that is lower than the CFCs it is intended to replace. HCFCs 123 and 124 and HFC-134a may also present a threat to ecosystems through their atmospheric breakdown product, trifluoroacetic acid (TFA). Under adverse conditions (arid climate, urban air pollution, high evaporation), TFA can accumulate in wetlands, potentially reaching levels that are toxic to plants (Tromp et al., 1995; Schwarzbach, 1995).

In refrigeration and air-conditioning applications, many new systems are expected to use HFC-134a, which has low toxicity and no ozone-depletion potential but has a significant global warming potential. HFC-134a is used in new automobile air conditioners, but retrofitting existing systems to use this compound can be expensive, depending on the automobile make and model year. Largely to satisfy demands of the automobile industry, Du Pont acceded to an EPA request that it produce its quota of CFC-12 in 1995 instead of the early phaseout it had pledged (*Du Pont Corporate News,* December 17, 1993).

HCFC-123 has low ozone depletion and global warming potentials (Table 3.1), but has caused benign tumors in rodent studies, leading to concern about its chronic toxicity. HCFC-22 and a variety of HCFC and HFC blends are also expected to be used in refrigeration and air-conditioning applications. Additional alternatives include ammonia and simple hydrocarbons that are flammable (such as propane, butane, and propylene). Ammonia also requires higher pressures (and consequently heavier compressors and tubing). Not-in-kind alternatives include replacing the vapor-compression cycle with evaporative-water cooling systems (used in some office buildings in the western United States) or, in niche applications, Stirling cycle systems (in which the refrigerant, typically helium, undergoes no phase transition).

Substitutes for CFCs in foam blowing depend on the type of foam. The EPA has approved the use of HCFC-141b, despite its relatively high ODP (0.1), for applications where low thermal conductivity is required, as with closed-cell insulating foams used in construction and refrigeration. In other applications (for example, cushioning foams), methylene chloride has long been used, despite its low exposure allowance in the workplace. Some of the newer HFCs are suitable for these applications, as are a number of hydrocarbons and carbon dioxide. With the exception of HFCs, many of these alternatives are volatile organic compounds and are subject to tropospheric ozone regulations. Because the blowing agent constitutes a relatively large share of production cost, final product costs are likely to increase by a larger percentage in foam than in other CFC applications.

A range of alternative cleaning technologies exist. In addition to "no-clean" solders and cutting oils, there are various aqueous and semi-aqueous cleaners, organic and chlorinated solvents, and supercritical fluid technologies. Substitutes are often specialized to application, and include a number of diverse blends. The organic solvents are typically flammable, and the chlorinated solvents (trichloroethylene, methylene chloride, and perchloroethylene) have low exposure allowances and are subject to hazardous-waste regulations. Perfluorocarbons (for example, perfluorobutane) are approved by the EPA for precision cleaning of electronic parts when other alternatives are infeasible, despite their extremely long atmospheric residence times (on the order of 10,000 years) and associated high global warming potentials.

In fire suppression systems, the leading substitutes for halons are conventional alternatives such as carbon dioxide, dry chemical, and foam extinguishants. When total-flooding systems are required or sensitive equipment is to be protected, alternatives include several HFCs (134a, 4310, and 23), and HCFCs 123 and 124 may be used as streaming agents.

| The Role of Risk Assessment

To date, substantial change in the U.S. CFC industry has occurred in two periods: the near total phaseout of CFC use in aerosols in the mid-1970s, and the beginning of an anticipated phaseout of most other applications under the Montreal Protocol and its amendments in the late 1980s and early 1990s. The roles of government and private action differed dramatically between these two periods. Moreover, the nature and extent of risk assessment also differed between the two periods, but in each case it appears to have been critical to the outcome.

In the aerosol phaseout, consumer pressure and the availability of low-cost substitutes led to rapid reduction of CFC use in aerosols. The regulation proscribing aerosol use beyond a list of "essential" applications codified a shift that had in large measure already occurred in the market. In contrast, the Montreal Protocol and its domestic implementing regulations have been a much more significant factor in driving reductions in CFC use in other applications, in large measure because the alternative technologies are less well developed and more expensive.

In the 1970s aerosol phaseout, formal risk assessment was limited. (This is perhaps not surprising, as the field of risk assessment was itself poorly developed at this time.) Physical scientists, beginning with Molina and Rowland, presented a persuasive theoretical argument that the continued release of CFCs to the atmosphere could lead to substantial reductions in stratospheric ozone, although no reduction had yet been observed. Through more established scientific studies, it was asserted that reductions in ozone would lead to increases in ultraviolet radiation flux at the earth's surface, and that such increases, particularly in the UV-B frequencies, would promote human skin cancers and would be harmful to sea life and crops. Changes in UV-B flux and effects on human health and other end points were not formally quantified, nor were possibilities for offsetting effects (increases in tropospheric ozone, for example) or adaptive behavior by humans or wildlife given much attention. Instead the analytic end point was equilibrium globally averaged ozone depletion, and judgments were implicitly made about the level of depletion that would be "acceptable."

In the United States, a credible scientific theory of how CFCs would deplete stratospheric ozone, without quantification of the ensuing consequences, was sufficient to induce the rapid phaseout of most aerosol applications. This theory may have been sufficient to induce the U.S. phaseout only because of the coexistence of several additional factors: 1) the virtually unprecedented global and potentially catastrophic nature of the threat; 2) the apparent ease of reducing emissions, by replacing what were perceived as frivolous uses in personal-care and other aerosol products with alternative propellants or dispensers that were reasonably satisfactory and, in many cases, less expensive; and 3) the contemporary enthusiasm for environmental protection, reflected in the celebration of the first Earth Day and the establishment of the EPA in 1970, and in the passage of major environmental legislation protecting air, water, soil, and endangered species.

When environmental risks are perceived as potentially catastrophic, and costs of change are perceived as low, a credible theory may be sufficient inducement for change.

This limited risk assessment was not sufficient to produce large changes in Europe or most other areas of the world (except for Canada, Norway, and Sweden, which also limited aerosol use). In part, the relatively limited response in the other major producing regions, Europe and Japan, may reflect a less aggressive approach to government regulation of industry, weaker environmental movements at the time, and, in Europe, a perceived connection between the (largely American) research describing the CFC-ozone link and the earlier American work linking supersonic transports to ozone depletion, which had been perceived as a threat to the joint British-French development of the Concorde. (As described above, the European Community did cap CFC production capacity and limit aerosol use in 1980, but neither the cap nor the aerosol limit was expected to bind in the near term, in part because aerosol use was declining in response to consumer preferences, as it had in the United States; Maxwell and Weiner, 1993: 23.)

Risk assessment in the 1980s, leading up to the Montreal Protocol and its amendments, has been much more complete, as represented by the 1987 EPA regulatory impact analysis. As in the earlier period, however, most of the attention has been directed to the physical science of ozone depletion, with much less emphasis on description and quantification of its consequences.

Observations of U.S. officials support the view that regulation — both the aerosol ban of the 1970s and the Montreal Protocol of the 1980s — was based primarily on the credible threat of global-scale damages, and not on comprehensive analysis of the marginal benefits and costs of incremental control options. Describing the aerosol ban, Food and Drug Administration commissioner Alexander Schmidt stated, "It's a simple case of negligible benefit measured against possible catastrophic risk, both for individual citizens and for society" (Dotto and Schiff, 1978: 286). Similarly, upon succeeding William Ruckelshaus as EPA administrator in January 1985, Lee Thomas was briefed on the CFC issue. He recalls, "I just took a black-and-white view when I saw the data. I knew we had to get [CFCs] out of process. It didn't appear that even a little bit of them was going to be safe" (Litfin, 1994: 73). Thomas's decision was not based solely on scientific assessment of the risk: few scientists offered policy prescriptions, and most who did thought that 50 percent emission reductions would be sufficient. In particular, Thomas's top science advisers did not advocate near-total phaseout of CFCs: Robert Watson of NASA testified to Congress that "the science doesn't justify a 95 percent cut" (Litfin, 1994: 103–104). Instead, Thomas's position was a risk management decision based on what would subsequently become known as the "precautionary principle;" in this, he differed from others in the administration. "[Presidential science adviser William] Graham looked at it from a purely scientific perspective, whereas I looked at it from a policy perspective. Where there was uncertainty, he thought we needed more research, and I thought we needed to be cautious. We just looked at the same thing and came to two different conclusions" (Litfin, 1994: 104).

Although analysis of the health and other consequences of ozone depletion was comparatively limited, the central link from CFC emissions to ozone depletion was extensively scrutinized. As described above, the National Academy of Sciences released a series of four reports between 1976 and 1984 assessing the then-current understanding of the CFC-ozone link. Although the anticipated magnitude of ozone depletion varied substantially between reports, the fundamental connection was consistently supported. In addition, a major assessment report sponsored by the World Meteorological Organization, the United Nations Environment Programme (UNEP), NASA, and several other organizations was released in 1986, further confirming the basic link. This report highlighted anticipated nonuniformities in ozone depletion, so that even if global depletion were modest, depletion in the temperate and higher latitudes, where most industrialized countries are located, could be much more significant. Despite fundamental support for the science, at the time the Montreal Protocol was signed, uncertainty about the magnitude of ozone depletion that would ensue in the absence of global regulation was daunting. Combining uncertainties about future CFC and halon emissions (Hammitt et al., 1987) with uncertainties about the atmospheric chemistry suggests an uncertainty range from about zero to 50 percent depletion over subsequent decades (Hammitt, 1995).

The 1988 Ozone Trends Panel report provided the first comprehensive evidence of observed as opposed to theoretical global ozone loss. This report enhanced the perception, forcibly generated by discovery of the Antarctic ozone hole, that scientists did not fully understand the ozone layer and that ozone losses were likely to exceed projections. As Robert Watson, chair of the Ozone Trends Panel, testified to Congress, "Our models are not doing a good job. . . . All of the previous reports have said there is no statistically significant trend since the 1960s. . . . What we are reporting is clearly a statistically meaningful decline in ozone" (Cogan, 1988: 67).

The combination of these national and international assessment reports, the discovery of the ozone hole, and the documentation of global ozone loss effectively limited the ability of industry and others to claim that the evidence remained inadequate to justify some response. In addition, the EPA provided quantitative estimates of the risks of ozone depletion in the regulatory impact analysis of its proposed rules implementing the Montreal Protocol restrictions, as required by Executive Order 12291. The monetized benefits of the regulation, $6.5 trillion, grossly exceeded the estimated costs, $27 billion (both figures are present values through 2075), making an apparently compelling case for regulation.

Ironically, at the same time that the EPA was providing quantitative estimates of the consequences and costs of ozone depletion, the primary end point used in policy discussion was retreating to an earlier link on the causal chain. Whereas equilibrium ozone depletion had been the primary analytic end point, attention now shifted to the atmospheric or stratospheric concentration of chlorine.[9] This shift was driven by persistent, if not growing, uncertainties in estimating the level of ozone depletion conditional on a specified atmospheric chlorine concentration. As noted above, neither the Antarctic ozone hole nor the levels of depletion reported by the Ozone Trends Panel were predicted by existing atmospheric models. In one expression of this view, Du Pont's briefing paper that accompanied its announcement of withdrawal from CFC production stated, "A prudent course of action would be to reduce atmospheric concentrations of CFCs to the levels that existed before the observed ozone decrease began to occur" (Cogan, 1988: 75). Subsequently, the science panel set up under the Montreal Protocol used this approach to conclude that, even with a global

[9] The EPA first presented an analysis using atmospheric chlorine as the end point at the September 1986 Leesburg, Virginia, scientific workshop it jointly sponsored with the UNEP.

phaseout of CFCs, halons, and some related compounds, the ozone hole will persist until about the middle of the twenty-first century, when stratospheric chlorine levels are projected to fall from their current level of about 3.3 to 2 parts per billion, the level at which the hole first appeared (WMO, 1994: 13.9). (This analysis assumes the reversibility of the processes producing the hole; Parson, 1993: 48–49). One effect of this change in perspective was to direct attention to much larger emission reductions, to prevent further increases in atmospheric chlorine that would occur even with the 50 percent emission reduction stipulated by the Montreal Protocol.

In addition to risk analysis, the perceived economic interests of various industrial firms appear to have played a role in shaping government regulation and industrial response. As noted above, Johnson Wax and other firms switched away from CFC aerosol propellants even before the U.S. ban, in order to satisfy consumers concerned about the link to ozone depletion. In the mid-1980s, Du Pont and other producers were united in their opposition to unilateral U.S. regulations, fearing further disadvantage vis-à-vis their international competitors. In addition, it is at least plausible that sophisticated producers like Du Pont and ICI (the dominant producer in the United Kingdom) believed they would benefit by government regulation of CFCs, as regulation would transform the market from selling CFCs, which had become low-margin commodity chemicals, to the development and marketing of specialized substitute chemicals in which these firms might expect to have a competitive advantage over smaller rivals. According to UNEP executive director Mostafa Tolba, "The difficulties in negotiating the Montreal Protocol had nothing to do with whether the environment was damaged or not. It was all who was going to gain an edge over whom; whether Du Pont would have an advantage over the European companies or not" (Maxwell and Weiner, 1993:32).

A somewhat puzzling aspect of this history is the extent to which policy analysis has focused on production and use, rather than emissions, of CFCs. Although the prospect of controlling emissions from millions of diverse sources is daunting, a risk-analytic perspective would direct attention to evaluating alternative means for reducing harm, such as 1) preventing atmospheric release of CFCs without necessarily preventing their use in applications like refrigeration, air conditioning, and cleaning, or 2) preventing ozone depletion once

CFCs were released to the atmosphere through some form of global environmental engineering (Cicerone, Elliott, and Turco, 1991, 1992; Amato, 1994). Similarly, it appears that relatively little attention was directed toward evaluating the risks of substitute processes or compounds until after the international decision to regulate CFCs was made. Since that time, substantial investments have been made by industry groups in toxicity testing and in evaluating the environmental effects of alternative compounds: about $20 million by PAFT and $10 million by the Alternative Fluorocarbons Environmental Acceptability Study (personal communication, T. Vogelsberg).

Most of the scientific research has been directed toward understanding the mechanisms controlling stratospheric-ozone concentrations. Research on subsequent links in the chain from CFC and halon emissions to human health and ecosystem effects has been much more limited in both scope and results, with funding typically less than 1 percent of the amount spent on atmospheric work. While the United States alone was spending on the order of $200 million per year on atmospheric research, the world as a whole was spending less than $1 million per year studying effects (Litfin, 1994: 55–56, 163). Similarly, there has been limited assessment of the likely level of future CFC emissions in the absence of additional regulations.

Much of the risk associated with ozone depletion is thought to be produced via increases in UV-B flux at the earth's surface. Yet despite evidence of ozone depletion, evidence of an increase in UV-B flux has been more difficult to obtain, as there are few long-term monitoring series and UV-B radiation is highly dependent on meteorological conditions, atmospheric haze, and other factors. As late as 1988, scientists found not an increase but rather an apparent decrease in ground-level UV-B (Scotto et al., 1988), most likely because of increasing tropospheric ozone. It was not until 1993 that ozone-related increases in ground-level UV-B flux were reported (Kerr and McElroy, 1993). That analysis employed data for a single city (Toronto) and consequently could not provide evidence of widespread changes in radiation, nor could it ensure that local factors were not dominant; indeed, UV-B radiation in Toronto subsequently decreased, suggesting the 1992–93 levels were a perturbation rather than a trend (WMO, 1994). Evidence of the effect of the Antarctic ozone hole on UV-B radiation and phytoplankton was first reported in 1992 (Smith et al., 1992).

The links from UV-B flux to skin cancer and other effects are also uncertain. Although there has been much speculation about effects of UV-B on incidence of frequently lethal melanoma, the occurrence of this cancer on body parts that are rarely exposed to sunlight suggests that other factors may be important. The links from chronic UV-B exposure to nonmelanoma skin cancers are more certain, but these cancers are (fortunately) readily treated with generally good results. An association between UV-B exposure and cataracts has been established for men, but not for women (Cruickshanks, Klein, and Klein, 1992).

Was the Environment Protected?

There is no doubt that production and use of CFCs and halons have declined, and the measured increase in atmospheric concentrations of these gases has slowed. On balance, the change is almost certainly good for human health and the environment, although the evidence for this is limited. Whether or not the internationally agreed emission reductions approximate an optimal path is less clear.

Much of the technological change has come in the form of source reduction. CFC aerosol propellants were replaced by hydrocarbon propellants and by alternative dispensers, such as pump sprays and roll-ons. Hydrocarbon propellants are flammable, producing an offsetting risk, although experience has not shown this risk to be large. Offsetting risks associated with alternative dispensers are presumably minimal.

Changes in nonaerosol applications have been dominated by the development and substitution of alternative compounds for refrigeration, foam blowing, and solvent applications. Some of the substitutes used in foam blowing and as solvents are more toxic to humans or contribute to photochemical smog (for example, methylene chloride). In refrigeration, the substitute HCFCs and HFCs are largely contained, so risks associated with their possible toxicity are small. Initially, there was concern that the newer compounds would require larger compressors and have poorer insulating properties, leading to higher energy use or requiring additional materials and volume to achieve comparable performance to CFC-based refrigeration systems, but recent studies suggest this effect is small or nonexistent (Fischer et al., 1991; Fischer, Tomlinson, and Hughes, 1994; Arthur D. Little, Inc., 1994). Although the substitute compounds generally have instantaneous effects on radiative forcing that are comparable to those of the

CFCs they replace, their typically shorter atmospheric lifetimes lead to reduced total contributions to climate change.

| Conclusions

Risk assessment has played a critical role in the processes that are leading to a virtually complete phaseout of a formerly ubiquitous class of industrial compounds. However, the risk assessment that was undertaken was very imbalanced in its emphasis: the vast majority of analytic resources and attention were directed to understanding the central link between CFC emissions and stratospheric ozone depletion, with much less attention paid to evaluating the health and other consequences of ozone depletion; the likely future CFC emissions, absent regulations; and the policy and technological measures and costs of either reducing CFC emissions or offsetting their effects.

In the early stages, a very limited risk assessment, essentially a hazard identification, was sufficient (in the United States and a few other countries) to drive CFCs from most aerosol applications, thereby reducing U.S. CFC use by roughly one-third. In later stages, when further reductions in CFC emissions required more costly changes, scientific assessment of the risk of continued CFC emissions remained the primary factor driving industrial change. In these later stages, more complete risk assessments estimating the magnitude of health effects and even their monetary costs were conducted, but these appear to have had modest effects at most on the policy process. Indeed, despite the quantification of health effects from ozone depletion, persistent uncertainty about the relationship between CFC emissions and ozone levels has led the policy discussion to focus its analysis on an earlier stage in the causal chain, atmospheric chlorine concentrations, and to make judgments about the acceptable chlorine level by reference to effects, such as the ozone hole, observed at chlorine levels lower than those existing today and likely to exist for several decades.

The relative unimportance given to quantified health and environmental risks appears to reflect the global, uncertain, and potentially catastrophic consequences of the threat. As was emphasized early in the debate, terrestrial life only developed after early photosynthetic organisms had produced oxygen, and with it, atmospheric ozone in sufficient quantity to screen out much of the ultraviolet radiation; humans and other terrestrial organisms could not be expected to survive without this shield. (Ironically, it was ultraviolet radiation that provided the energy necessary to synthesize the initial organic compounds that produced life on earth; Dotto and Schiff, 1978:27.) Thus, compelling evidence supporting the CFC-ozone link was decisive in leading to regulation and economic shifts. The attitude prevailing at the 1989 Helsinki meeting of the parties to the Montreal Protocol was the attitude that led to the negotiation of the protocol itself: "Don't fool with Mother Nature" (Litfin, 1994: 139).

| Sources

Amato, I., 1994. "A High-Flying Fix for Ozone Loss," *Science* 264:1401–1402.

Arthur D. Little, Inc., 1994. *Update on Comparison of Global Warming: Implications of Cleaning Technologies Using a Systems Approach,* Report to AFEAS, Reference 46342 (Cambridge, Mass.: Arthur D. Little, Inc.).

Alternative Flurocarbons Environmental Acceptability Study (AFEAS), 1996. *Production, Sales and Atmospheric Release of Fluorocarbons through 1994* (Washington, D.C.: AFEAS).

Benedick, R. E., 1991. *Ozone Diplomacy: New Directions in Safeguarding the Planet* (Cambridge: Harvard University Press).

Brodeur, P., 1975. "Annals of Chemistry: Inert," *The New Yorker,* April 7.

———, 1986. "Annals of Chemistry: In the Face of Doubt," *The New Yorker,* June 9.

Cagin, S., and P. Dray, 1993. *Between Earth and Sky: How CFCs Changed Our World and Endangered the Ozone Layer* (New York: Pantheon).

Chemical Manufacturers Association (CMA), 1986. *Production, Sales, and Calculated Release of CFC-11 and CFC-12 through 1985* (Washington, D.C.: Chemical Manufacturers Association).

Cicerone, R. J., S. Elliott, and R. P. Turco, 1991. "Reduced Antarctic Ozone Depletions in a Model with Hydrocarbon Injections," *Science* 254:1191–1194.

———, 1992. "Global Environmental Engineering," *Nature* 356:472.

Cicerone, R. J., R. S. Stolarski, and S. Walters, 1974. "Stratospheric Ozone Destruction by Man-Made Chlorofluoromethanes," *Science* 185:1165–1167.

Cogan, D. G., 1988. *Stones in a Glass House* (Washington, D.C.: Investor Responsibility Research Center).

Connell, P. S., 1986. *A Parameterized Numerical Fit to Total Column Ozone Changes Calculated by the LLNL 1D Model of the Troposphere and Stratosphere,* Lawrence Livermore National Laboratory UCID-20762 (Livermore, Calif.: LLNL).

Cook, E., 1996. "Marking a Milestone in Ozone Protection: Learning from the CFC Phase-Out," *WRI Issues and Ideas,* January (Washington, D.C.: World Resources Institute).

Cruickshanks, K. J., B. E. K. Klein, and R. Klein, 1992. "Ultraviolet Light Exposure and Lens Opacities: The Beaver Dam Eye Study," *American Journal of Public Health* 82:1658–1662.

Crutzen, P. J., 1970. "The Influence of Nitrogen Oxides on the Atmospheric Ozone Content," *Quarterly Journal of the Royal Meteorological Society* 96:320–325.

Daniel, J. S., S. Solomon, and D. L. Albritton, 1994. "On the Evaluation of Halocarbon Radiative Forcing and Global Warming Potentials," *Journal of Geophysical Research.*

Dotto, L., and H. Schiff, 1978. *The Ozone War* (Garden City, N.Y.: Doubleday).

Environmental Protection Agency (EPA), 1980. "Ozone-Depleting Chlorofluorocarbons: Proposed Production Restriction," *Federal Register* 45:66726, October 7.

———, 1987a. "Protection of Stratospheric Ozone," *Federal Register* 52:47489–47523, December 14.

———, 1987b. *Regulatory Impact Analysis: Protection of Stratospheric Ozone,* Stratospheric Protection Program, Office of Program Development, Office of Air and Radiation, December (Washington, D.C.: Environmental Protection Agency).

———, 1994. "Protection of Stratospheric Ozone," *Federal Register* 59:13044–13146, March 18.

Farman, J. C., B. G. Gardiner, and J. D. Shanklin, 1985. "Large Losses of Total Ozone in Antarctica Reveal Seasonal $C1O_x/NO_x$ Interaction," *Nature* 315:207–210.

Fischer, S. K., P. J. Hughes, P. D. Fairchild, et al., 1991. *Energy and Global Warming Impacts of CFC Alternative Technologies,* Report to AFEAS and the U.S. Department of Energy, Oak Ridge, Tennessee, Oak Ridge National Laboratory, and Arthur D. Little, Inc., Cambridge, Massachusetts.

Fischer, S. K., J. J. Tomlinson, and P. J. Hughes, 1994. *Energy and Global Warming Impacts of Not-in-Kind and Next Generation CFC and HCFC Alternatives,* Report to AFEAS and the U.S. Department of Energy (Oak Ridge, Tennessee: Oak Ridge National Laboratory).

Fisher, D. A., C. H. Hales, D. L. Filkin, et al., 1990. "Model Calculations of the Relative Effects of CFCs and Their Replacements on Stratospheric Ozone," *Nature* 344:508–512.

Fisher, D. E., 1990. *Fire and Ice: The Greenhouse Effect, Ozone Depletion, and Nuclear Winter* (New York: Harper and Row).

Hammitt, J. K., 1995. "Outcome and Value Uncertainties in Global-Change Policy," *Climatic Change* 30:125–145.

———, in press. "Stratospheric-Ozone Depletion," in R. D. Morgenstern (ed.), *Economic Analysis at EPA: Assessing Regulatory Impacts* (Washington, D.C.: Resources for the Future).

Hammitt, J. K., F. Camm, P. S. Connell, et al., 1987. "Future Emission Scenarios for Chemicals That May Deplete Stratospheric Ozone," *Nature* 330:711–716.

Hammitt, J. K., K. Wolf, F. Camm, et al., 1986. *Product Uses and Market Trends for Potential Ozone-Depleting Substances, 1985–2000* (Santa Monica, Calif.: Rand).

Intergovernmental Panel on Climate Change (IPCC), 1994. *Radiative Forcing of*

Climate Change (Geneva: World Meteorological Association/UN Environment Programme).
Johnston, H., 1971. "Reduction of Stratospheric Ozone by Nitrogen Oxide Catalysts from Supersonic Transport Exhaust," *Science* 173:517–522.
Kerr, J. B., and C. T. McElroy, 1993. "Evidence for Large upward Trends of Ultraviolet-B Radiation Linked to Ozone Depletion," *Science* 262:1032–1034.
Litfin, K. T., 1994. *Ozone Discourses: Science and Politics in Global Environmental Cooperation* (New York: Columbia University Press).
Lovelock, J. E., R. S. Maggs, and R. J. Wade, 1973. "Halogenated Hydrocarbons in and over the Atlantic," *Nature* 241: 194–196.
Maxwell, J. H., and S. T. Weiner, 1993. "Green Consciousness or Dollar Diplomacy? The British Response to the Threat of Ozone Depletion," *International Environmental Affairs* 5:19–41.
Molina, M. J., and R. F. Rowland, 1974. "Stratospheric Sink for Chlorofluoromethanes: Chlorine Atom Catalysed Destruction of Ozone," *Nature* 249:810–811.
Molina, M. J., L. L. Tso, L. T. Molina, and F. C. Y. Wang, 1987. "Antarctic Stratospheric Chemistry of Chlorine Nitrate, Hydrogen Chloride and Ice: Release of Active Chlorine," *Science* 238:1253–1258.
National Academy of Sciences (NAS), 1976. *Halocarbons: Effects on Stratospheric Ozone* (Washington, D.C.: National Academy Press).
———, 1979. *Protection against Depletion of Stratospheric Ozone by Chlorofluorocarbons* (Washington, D.C.: National Academy Press).
———, 1982. *Causes and Effects of Stratospheric Ozone Depletion: An Update* (Washington, D.C.: National Academy Press).
———, 1984. *Causes and Effects of Stratospheric Ozone Depletion: Update 1983* (Washington, D.C.: National Academy Press).
National Aeronautics and Space Administration (NASA), 1988. *Executive Summary, Ozone Trends Panel* (Washington, D.C.: NASA).
National Research Council (NRC), 1975. *Environmental Impact of Stratospheric Flight: Biological and Climatic Effects of Aircraft Emissions in the Stratosphere* (Washington, D.C.: National Academy of Sciences).
Parson, E. A., 1993. "Protecting the Ozone Layer," in P. M. Haas, R. O. Keohane, and M. A. Levy, eds., *Institutions for the Earth: Sources of Effective International Environmental Protection* (Cambridge: Harvard University Press).
Pukelsheim, F., 1990. "Robustness of Statistical Gossip and the Antarctic Ozone Hole," *The IMS Bulletin* 19:540–542.
Roan, S., 1989. *Ozone Crisis: The 15-Year Evolution of a Sudden Global Emergency* (New York: John Wiley and Sons).
Schwarzbach, S. E., 1995. "CFC Alternatives under a Cloud," *Nature* 376:297–298.
Scotto, J., G. Cotton, F. Urbach, et al., 1988. "Biologically Effective Ultraviolet Radiation: Surface Measurements in the United States, 1974 to 1985," *Science* 239:762–764.
Smith, R. C., B. B. Prezelin, K. S. Baker, et al., 1992. "Ozone Depletion: Ultraviolet

Radiation and Phytoplankton Biology in Antarctic Waters," *Science* 255:952–959.

Solomon, S., R. R. Garcia, F. S. Rowland, and D. J. Wuebbles, 1986. "On the Depletion of Antarctic Ozone," *Nature* 321:755–758.

Somerville, R. C. J., 1996. *The Forgiving Air: Understanding Environmental Change* (Berkeley: University of California Press).

Stoel, T. B. Jr., A. S. Miller, and B. Milroy, 1980. *Fluorocarbon Regulation: An International Comparison* (Lexington, Mass.: Lexington Books).

Stolarski, R. S., and R. J. Cicerone, 1974. "Stratospheric Chlorine: A Possible Sink for Ozone," *Canadian Journal of Chemistry* 52:1610–1615.

Tromp, T. K., M. K. W. Ko, J. M. Rodriguez, and N. D. Sze, 1995. "Potential Accumulation of a CFC-Replacement Degradation Product in Seasonal Wetlands," *Nature* 376:327–330.

World Meteorological Organization (WMO), 1985. *Atmospheric Ozone 1985,* Global Ozone Research and Monitoring Project, publication 16.

———, 1994. *Scientific Assessment of Ozone Depletion: 1994,* Global Ozone Research and Monitoring Project, publication 37.

Wuebbles, D. J., 1981. *The Relative Efficiency of a number of Halocarbons for Destroying Stratospheric Ozone,* Lawrence Livermore National Laboratory UCID-18924 (Livermore, Calif.: LLNL).

———, 1983. "Chlorocarbon Emission Scenarios: Potential Impact on Stratospheric Ozone," *Journal of Geophysical Research* 88(C2):1433–1443.

Wuebbles, D. J., F. M. Luther, and J. E. Penner, 1983. "Effect of Coupled Anthropogenic Perturbations on Stratospheric Ozone," *Journal of Geophysical Research* 88(C2):1444–1456.

FOUR

Cleaning Up Dry Cleaners

Kimberly M. Thompson

During the past two decades, the dry cleaning industry has been regulated under several environmental statutes and has achieved a substantial amount of emission control. Currently, this multibillion dollar industry cleans more than 600 million kilograms of clothes per year (a typical man's or woman's suit weighs about 1 kg) in more than 30,000 individual facilities in the United States (EPA, 1991a). Approximately 85 to 90 percent of the industry currently uses perchloroethylene (perc) as a primary solvent (EPA, 1991a; CEC, 1992), with most of the remainder of the dry cleaners (predominately in the Southeast) using petroleum solvents (called Stoddard solvents) where fire codes permit their use (Wolf and Myers, 1987). Most federal regulations have targeted facilities using perc, and consequently this chapter focuses them.

Dry cleaners became the target for environmental health regulations in the late 1970s, when the Environmental Protection Agency began requiring reductions in emissions of volatile organic compounds (VOCs) to reduce smog and when the results from animal tests first led to concern that perc might be a human carcinogen. Regulations have been based largely on qualitative assessments of perc in keeping with the statutory requirements, but some quantitative risk assessments and cost-benefit analyses have also been performed. In spite of years of study, federal agencies have experienced difficulties classifying perc as a carcinogen. Overall, the regulatory focus in risk assessment has been mainly on the issue of whether or not perc is a human carcinogen (corresponding to the hazard identification step in a risk assessment), and less effort has been devoted to characterizing risks (corresponding to the last three steps of a risk assessment).

Examination of the regulatory assessment processes for perc provides perspective on the effects of requiring quantitative risk assessment and cost-benefit analysis as prerequisites to emissions control, and on the substantial resource expenditures required when the scientific findings are equivocal. Given the limited amount of resources available, this case provides evidence for examining the question; Have society's resources been well spent? In addition, by considering the response of a single industry to multiple regulations, this case facilitates comparisons of the role of risk assessment under different regulatory frameworks. Overall, emissions from dry cleaners have declined and the industry continues to adopt pollution prevention measures. It appears that qualitative concerns about danger played a larger role than quantitative analysis in this reduction, but the threat of quantitative risk results has also helped to motivate the industry to change.

For perspective, this chapter begins with a description of the evolution of the U.S. dry cleaning industry, a discussion of the industry's choice of perc as its primary solvent, and a description of the types of perc releases from dry cleaners. The chapter then describes the evidence for perc's carcinogenicity and the difficulties agencies have had in trying to classify perc as a human carcinogen. Next, the chapter describes the development of key regulations that have led to emissions reductions, and demonstrates the combined effects of the regulations in making the industry reduce its emissions of perc to the environment. The last part of this chapter explores the role of risk assessment in these regulations and speculates on the future role of risk assessment for the industry.

| Evolution of Dry Cleaning

Recognition of the importance of cleaning clothes can be traced back to biblical times. In the Old Testament, Leviticus chapters 11, 13, 14, 15, and 17 all indicate circumstances under which a person's clothes must be washed before that person may be considered clean. The cleaning process entails the transfer of contaminants from the clothes to a solvent. For centuries, water served as the only solvent available for textile cleaning. Treatises providing formulas for cleaning all kinds of textiles with water date back at least to 1707 (Johnson, 1971). Perhaps one of the most famous and exhaustive of these treatises is Thomas Love's 1854 handbook entitled, *The Art of Dyeing, Cleaning, Scouring, and Finishing.*

The Romans probably invented the first dry cleaning process by using absorbent clays and powders (for example, Fuller's earth) to remove grease stains from clothes (Johnson, 1971). The term "dry" cleaning originated as a reference to the use of dry powders and absorbent clays for cleaning. The process of cleaning clothes using an organic solvent (instead of water) also became known as dry cleaning because it did not involve the use of water, and current use of the term refers solely to this process.

Dry cleaning with organic solvents became possible during the Industrial Revolution. At that time, inexpensive organic solvents became available as by-products from other industries (for example, benzene came from coal tar produced by the coal gas industry). Jean-Baptiste Jolly of the French firm Jolly-Belin is credited with the first use of organic solvents for textile cleaning, in the 1820s (Johnson, 1971). In 1869, a Scottish firm called Pullars of Perth, introduced power-driven machinery for dry cleaning with benzene as the solvent that facilitated the development of dry cleaning as an industry.

Since the 1850s, the solvents and technologies used by the U.S. dry cleaning industry have changed dramatically. Table 4.1 lists different dry cleaning solvents, some of their properties, and approximate dates of introduction and periods of use. In the early 1900s, dry cleaning in the United States was conducted in large, centralized factories, and consumers typically had to wait at least a week for garments to be returned (Johnson, 1971). Dry cleaning machines used flammable petroleum solvents, and there were separate units for washing and drying (similar to domestic laundering machines), which allowed the hot dryer to be located away from the solvent tanks and the washer. The possibility of explosion and fire from dry cleaners (as well as many other industries) led cities to enact fire codes that prevented dry cleaners from operating in residential areas (Street, 1926). In 1928 the industry adopted standard specifications for dry cleaning solvents, and within a decade these "Stoddard solvents" became the industry's preferred solvents, because they were less flammable (and consequently safer) than other petroleum solvents (Lyle, 1977).

TABLE 4.1 | Dry cleaning solvents

SOLVENT	CLEANING ABILITY[a]	CONCERNS AND COSTS[b]	WHEN INTRODUCED	PRIMARY PERIOD OF USE BY U.S. INDUSTRY
Benzene (from coal tar)	Poor	F, V, T (human carcinogen)	1820s	1820–1860
Camphene (from turpentine)	Poor	F, V, T (potentiates cholesterol)	1820s	1820–1860
Petroleum solvents called "gasoline," and Stoddard solvents	Modest, KB = 30–40	F, V, T (suspected animal carcinogen), $1.20/gal	1860s	1860s to 1960s (some current use)
Carbon tetrachloride	Good, dries quickly	V, T (produces acute central nervous system depression, chronic liver injury; suspected animal carcinogen)	1920s–1930s in Germany	Early 1940s
Trichloroethylene	Good, KB = 124	D, S, V, T (suspected animal carcinogen)	1920s–1930s in Germany	Middle 1940s
Perchloroethylene	Good, KB = 92	V, T (suspected animal carcinogen), $4.00/gal	1920s–1930s in Germany	1940 to present
Chlorofluorocarbon-113	Modest, KB = 31	O, $16.00/gal	1968	Limited use since introduction to 1996
1,1,1-trichloroethane	KB = 130	D, S, O, T (suspected animal carcinogen), $6.00/gal	1985	Limited use since introduction to 1996

a. Relative cleaning abilities given in this column. KB represent the kauri-butanol value, which describes the degreasing ability of a solvent. A good dry cleaning solvent typically has a KB value between 30 and 100. Solvents with KB values lower than 30 are typically poor cleaners, while those with KB values above 100 tend to dissolve dyes. KB values from Wolf and Myers (1987).

b. Concerns and costs: F = solvent is flammable (solvents without Fs are nonflammable); T = possible toxicologic effects; D = solvent damages some clothes; S = solvent is relatively unstable and consequently may have to be replaced more frequently; O = solvent depletes stratospheric ozone; V = solvent is a volatile organic compound. Cost estimates are approximate (1985 dollars).

In the 1920s and 1930s, the Germans introduced three chlorinated solvents into dry cleaning: trichloroethylene, carbon tetrachloride, and perc (Johnson, 1971). Following World War II, Americans gained an increased sense of mobility, and at the same time good grooming became very important. These factors created demand for quick dry cleaning services and led the industry into a gradual decentralization and a shift to chlorinated solvents. In the late 1940s some small tailors began to perform dry cleaning in their existing urban and suburban facilities using nonflammable carbon tetrachloride. These businesses were able to provide convenient and fast (even same-day) cleaning service, because carbon tetrachloride is highly volatile and the clothes dried relatively quickly, without the use of a dryer. The drawback to using carbon tetrachloride for dry cleaning was the solvent's acute toxicity. Dry cleaners using carbon tetrachloride decided it was too dangerous because it made them feel ill, and they started to use trichloroethylene instead (Seitz, 1994). The use of trichloroethylene was relatively short-lived, however, because it made the dyes in some fabrics run (Seitz, 1994). The next alternative solvent to be adopted was perc, which turned out to be a good solvent for cleaning. Since perc was less volatile and more expensive than carbon tetrachloride, dry cleaners using perc also used dryers to decrease their servicing time and solvent costs.

| The Rise of Perc

Since the 1950s, dry cleaners in the U.S. gradually shifted from Stoddard solvents to perc as additional and more restrictive local fire codes were established. By 1970, over 50 percent of the industry used perc as its primary solvent (Johnson, 1971). Decentralization of the industry peaked in the late 1970s and early 1980s with the popularity of self-service coin-operated dry cleaning machines using perc that allowed consumers to dry clean their own clothes in laundromats. The industry currently appears to be undergoing a recentralization trend, with a resurgence of drop-off shops and the recent decline and near extinction of coin-operated facilities. Well over 90 percent of the existing dry cleaning facilities are classified as small businesses (EPA, 1991b) and solvent costs represent less than 1 percent of variable costs for a typical dry cleaner (Seitz, 1994).

Machines that perform both the washing and drying in a single unit were designed by the Germans in the 1930s, because the chlorinated solvents were considered to be too expensive to waste, to have an unpleasant odor, and to be unhealthy if too much solvent was inhaled (particularly carbon tetrachloride) (DOL, 1938). In spite of this invention, U.S. dry cleaners preferred machines with separate washers and dryers (called transfer machines), because facilities could operate the washer and dryer simultaneously and thus clean almost twice as much clothing in the same amount of time (or use smaller machines to clean the same amount of clothing). The single-unit machines (called dry-to-dry machines) did not gain wide acceptance in the United States until the 1960s when dry cleaners began replacing old equipment with dry-to-dry machines. Since the lifetime of machines is approximately twenty years, the shift has been gradual. In 1990, approximately two-thirds of all perc machines were dry-to-dry machines (EPA, 1991b).

In 1964 E. I. Du Pont de Nemours and Company (Du Pont) introduced chlorofluorocarbon-113 (CFC-113) as a dry cleaning solvent under the trade name Valclene. Because of its relatively high cost, only a small percentage of the U.S. industry (less than 1 percent) adopted this new technology. The CFC-113 systems became more popular in Australia, where they were used by up to 30 percent of dry cleaners (Seitz, 1994). In 1985, Dow Chemical introduced 1,1,1-trichloroethane (TCA), which was similarly adopted by only a very small number of U.S. dry cleaners. Due to their ability to deplete stratospheric ozone, production of both CFC-113 and TCA has been banned, and consequently they have been phased out of use as dry cleaning solvents (see Chapter 3). More recently, improved hydrocarbon solvents have also been introduced, but flammability remains an issue. Thus, perc remains the best-suited organic solvent for dry cleaning, due to its nonflammability, low acute toxicity, and efficacious cleaning. Recent initiatives to find other dry cleaning solvents have come full circle and focus on using water for cleaning clothes that are typically dry cleaned ("wet cleaning").

| Perc Releases from Dry Cleaners

Due to the nature of the dry cleaning process, facilities may release perc to the environment in a number of ways, as shown in Figure 4.1. Unlike domestic washers and dryers that use the wash solvent water

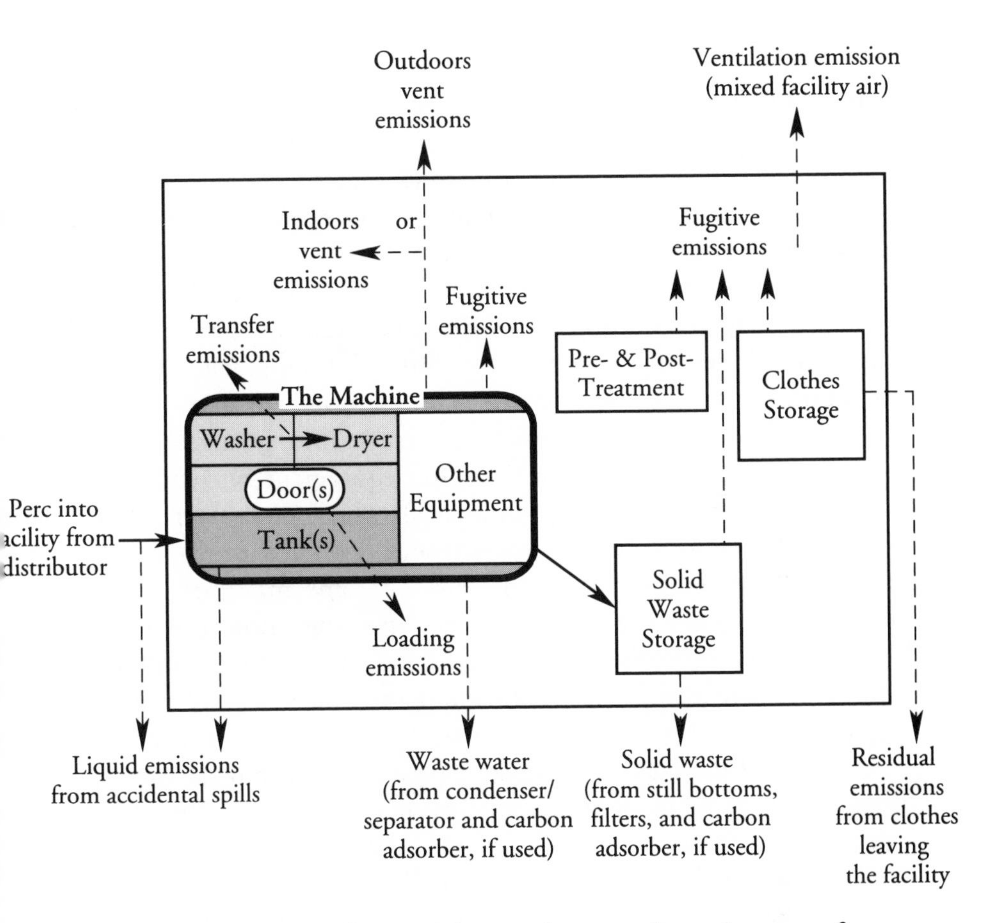

FIGURE 4.1 | Schematic of potential perc releases to the environment from a dry cleaner.

once and then discard it, dry cleaning equipment is designed to recycle perc. In order to reuse the perc, contaminants must be removed from it between washings. One way to remove contaminants is to pass the solvent through a series of filters. Depending on the type of filter used, residual contaminants and solvent collected on the filters may be placed in a still that is used to recover more perc. Used filters and still bottoms contain residual levels of perc, and consequently they constitute possible sources of solid waste. Some facilities add steam to the still because it makes the solvent boil at a lower temperature and consequently requires less energy. Adding steam to the still typically leads to the generation of some perc-contaminated waste water.

Perc is also recovered in the dryer and recycled back into the washer. Drying the clothes removes perc from the clothing by evaporation, although some residual levels may remain on the clothes. The evaporated perc vapor is sent to a condenser, where perc and water vapor condense into liquid. The liquid perc is separated from the perc-contaminated waste water in a separator. Air coming out of the condenser is typically recycled to the dryer. At the end of the drying cycle, the air in the machine may be vented out as fresh air from the shop is brought into the dryer to freshen the clothes (a process called aeration). In addition, the machine may draw in air from the shop when the machine door is opened and then send it out through a vent.

Perc vapors may escape from the equipment into the facility through leaks (called fugitive emissions), when the machine door is open, while clothes are transferred from the washer to the dryer for transfer machines, and through vents on some machines. Efficient control devices exist to recover perc in air vented from the machine (such as carbon adsorbers and refrigerated condensers). In addition to substantially reducing consumption of perc by the machine, using such devices may slightly increase the amount of solid waste or waste water generated by the facility. Ultimately perc released into the facility is released to the ambient air through normal ventilation of the facility and then out through open doors and windows. Finally, accidental perc spills may result in releases to the ground and air.

In the mid-1980s, it was estimated that most of the perc emitted by dry cleaners (approximately 80 to 90 percent) went into the ambient air, with most of the remaining 10 to 20 percent being emitted in the form of solid waste (Wolf and Myers, 1987). Amounts of perc released in waste water and on clothes are very small by comparison (less than 1 percent), and although the quantity of perc lost due to spills is unknown, it is also expected to be relatively small. Currently, the amount of perc emitted to the air by a single dry cleaning facility ranges from 40 to 80 percent, depending on the type of equipment used.

| Cancer Evidence

Releases of perc have been a regulatory concern because they may cause adverse health effects (ATSDR, 1993; CARB, 1991; EPA, 1985a; EPA, 1986a; EPA, 1991c). Reported effects range from acute central nervous system, liver, and kidney effects for exposure to relatively high levels of perc (those over 678,000 $\mu g/m^3$ or equivalently 100 parts per million) to recently reported induced color vision loss (Cavalleri et al., 1994; Echeverria, White, and Sampaio, 1995) and chronic kidney toxicity (Mutti et al., 1992; Price et al., 1995) at lower levels. However, at ambient and current occupational concentrations, carcinogenicity is of concern and it has been the primary basis for federal regulation to date.

In the 1970s several animal bioassays were conducted to assess perc's ability to cause cancer. Dow Chemical performed a study that exposed rats to perc in air at levels three and six times the permissible exposure limit (PEL) of 100 parts per million (ppm) over a period of twelve months to simulate occupational exposure. No statistically significant increases in cancer rates were observed (Rampy et al., 1978), although the EPA considered the study too short to be an adequate cancer study (EPA, 1985a). Two other limited studies of perc also failed to show evidence of a statistically significant increase in tumors in test animals (Theiss et al., 1977; Van Duuren et al., 1979).

But not all of the animal bioassays were negative. In 1974 the National Cancer Institute (NCI) initiated a two-year animal bioassay for perc that involved feeding perc in corn oil by gavage to both sexes of rats and mice. The study results, released in 1977, suggested that perc could cause cancer in mice (NCI, 1977). The mice exposed to perc exhibited a significantly higher number of hepatocellular carcinomas (malignant liver tumors) than the control mice. The survival of the rats in the NCI study was inadequate to support any conclusions about perc's carcinogenicity in rats. In 1986 the results from a two-year inhalation bioassay of perc conducted by the National Toxicology Program (NTP) were released (NTP, 1986). In the NTP study, dose-response trends were found for hepatocellular carcinomas and combined hepatocellular adenomas and carcinomas in mice, and for leukemias in rats. The exposed rats also exhibited higher numbers of kidney tumors, but the dose-response trend was not statistically significant.

The results of genotoxicity and mutagenicity studies for perc and its possible metabolites have been relatively unhelpful in efforts to classify perc (EPA, 1991c). Metabolism of perc by both rodents and humans may occur via two pathways to form potentially mutagenic or genotoxic metabolites: 1) oxidation of the double bond by the cytochrome P-450 system, and 2) conjugation of perc with glutathione. An EPA report concludes that

> The available data indicate that metabolism is a prerequisite for perchloroethylene mutagenicity. The data do not support classifying the parent compound per se as a mutagen. Although certain metabolites of oxidative metabolism may be mutagenic (e.g., the chloroacetaldehydes including chloral hydrate), these positive data are predominantly limited to in vitro studies. Moreover, perchloroethylene was assayed in the presence of several types of metabolic activation systems (e.g., liver homogenates and intact hepatocytes) that would favor oxidative metabolism and, under these conditions, predominantly negative results were found. Perchloroethylene may also be activated by a minor pathway involving conjugation with glutathione followed by renal processing of the S-conjugate. This S-conjugate is a beta-lyase dependent mutagen in the Salmonella/mammalian microsome assay. Mutagenic metabolites formed in the kidney could conceivably contribute to the tumors observed in male rat kidneys. However, these mutagenicity studies of perchloroethylene metabolites formed by the kidney are in vitro only (EPA, 1991c: 21).

Epidemiological evidence of perc's carcinogenicity has similarly been inconclusive (EPA, 1985a; CARB, 1991). Various studies of perc exposure report excess mortality from leukemia and from bladder, lung, cervix, kidney, skin, esophageal, and colon cancer (Blair, Decoufle, and Grauman 1979; Kaplan, 1980; Katz and Jowett, 1981; Lin and Kessler, 1981; Duh and Asal, 1984; Brown and Kaplan, 1985; Aschengrau et al., 1993; and Ruder, Ward, and Brown, 1994). Unfortunately, these studies suffer from a variety of problems, including potential confounding by smoking, exposure to other agents (like petroleum solvents), and socioeconomic status; possible misclassification of exposures (due to the inability to distinguish laundry workers from dry cleaning workers); and statistical instability due to the small numbers of deaths for specific types of tumors (EPA, 1985a; CARB, 1991). Consequently, efforts to understand and characterize perc's carcinogenicity have focused largely on the animal evidence.

The NCI results initiated a long, contentious, and on-going dispute between regulatory agencies and perc producers and users (including the dry cleaning industry) about the relevance of animal bioassay results to humans. A key question was: Should perc be formally classified as a potential human carcinogen?

| Carcinogen or Not?

Classifying perc as a carcinogen has proven to be remarkably difficult and controversial for each agency that has attempted to classify it.[1] At the EPA, the process has included numerous iterations between staff and the EPA's Science Advisory Board (SAB), and perc is one of the compounds that recently motivated the EPA to propose a new classification system (EPA, 1994a; EPA, 1996a).

In the 1970s, when animal tests were initiated and federal cancer policies were being developed, there was a sense that only a few of the substances tested would cause cancer in animals and would consequently be suspected as human carcinogens. In June 1978 the Consumer Product Safety Commission (CPSC) released its policy on classifying carcinogens, and it conditionally labeled perc as a "category A" human carcinogen, based on the results from the NCI study (CMR, 1978). Six producers of perc and the Louisiana Chemical Association (with support from the dry cleaning industry) sued the CPSC, asserting that the classification action was "arbitrary, capricious and unsupported by the record" (*Dow Chemical, USA v. CPSC*, 459 F. Supp. [W.D.La. 1978]: 378). The CPSC's cancer policy was overturned by the federal court because the CPSC had failed to provide opportunity for public comment as required by the Administrative Procedure Act, and the CPSC withdrew its classification of perc. Although the judge did not resolve the issue of how perc should be classified, the court's decision appears to have stopped further classification of and regulatory action on perc at the CPSC, and it may have influenced CPSC's strategy for addressing chronic hazards.

The EPA's regulation of perc as a potential carcinogen began in 1979 when its Carcinogen Assessment Group (CAG) published a draft health assessment document (HAD) for perc. Based on the NCI data and its interim cancer risk assessment guidelines, the CAG concluded that "substantial" evidence of human carcinogenicity existed for perc. Since the HAD was likely to form the basis for regulations, it was reviewed by an EPA Science Advisory Board (SAB) subcommittee in

[1] This section relies on Drye (1991), which reviews the carcinogen classification difficulties through 1990.

September 1980. It was commonly perceived that the board's approval of a health assessment document was expected prior to the use of the document's conclusions to support regulatory decision making, and that a substance's carcinogenic classification would influence how its use was regulated.

The SAB review provided an opportunity for perc producers and users to become involved in the EPA's process of evaluating the carcinogenicity evidence for perc. The Halogenated Solvents Industry Alliance (HSIA) and members of dry cleaning trade associations invested considerable resources in the review process. While the influence of any comments provided to the SAB is unknown, the board concluded that the Carcinogen Assessment Group's finding of "substantial evidence" was not supported by the health assessment document, and it asserted that the relevant question in the case of perc was whether the document included any evidence of human carcinogenicity for perc at all. The board questioned whether a stabilizer present in the perc used for the NCI bioassay may have caused the excess mouse liver tumors, and whether the mouse liver tumors were relevant to humans.

The relevance of the mouse liver tumors has been an area of contention for many chemicals that have been tested. Some scientists attribute the induction of liver tumors in the species of mice to the combination of recurrent liver damage that results from the high doses of the chemicals and rapid turnover of liver cells in the setting of a high background rate of liver tumors. The EPA maintains that the induction of mouse liver tumors can, in some cases, provide sufficient evidence of carcinogenicity, and it listed explicit criteria for when mouse liver tumors can be discounted in its 1986 cancer guidelines (EPA, 1986b).

The EPA predicted that rejection of the HAD and the proposed classification of perc would directly affect the development of regulatory action. Roy Albert, the EPA's acting director of the Carcinogen Assessment Group, told the board, "You may think this is casual advice to CAG, but the regulatory people can't ignore what you say. If you say a chemical is less than 'substantial' or 'suggestive,' it will stop a standard cold" (CRR, 1980). This may have been the case in the development of a national rule to reduce air emissions of perc, discussed below.

In 1982 the EPA released a new draft of the health assessment document that used the official classification scheme of the International Agency for Research on Cancer (IARC). The CAG concluded in this draft that perc was a "potential human carcinogen," and that according to the IARC classification scheme, the evidence for perc was "limited" and insufficient to "provide a firm conclusion on its carcinogenic potential in humans" (Drye, 1991: 102). In this round of review, the SAB criticized the ambiguity of the document's conclusions, the assumptions and modeling approach used by the agency to estimate a unit risk factor for perc, and the lack of context provided for interpretation of the unit risk. The unit risk is characterized as the upper-bound lifetime probability of cancer per unit of concentration of the substance in air, and it is based on the assumption that any finite dose could lead to a case of cancer. The agency policy at the time was to derive unit risks for compounds considered likely to be "carcinogens" (that is, those in category A or B).

In 1983 the EPA released another draft of the health assessment document. In this revision, the EPA characterized the animal evidence as "limited" and the epidemiological evidence as "inconclusive," corresponding to IARC carcinogen category 3 (not classifiable). However, in March 1984 the EPA modified its characterization of the animal evidence to be "nearly sufficient" and indicated that perc fell "in IARC category 3, but is close to 2b, i.e., the more conservative scientific view would regard [perc] as being close to a probable human carcinogen" (EPA, 1984a). This modification was made two months before the board reviewed the draft health assessment document and shortly after the Office of Drinking Water released a document that pushed for the regulation of perc as a carcinogen (EPA, 1984b).

During this round of review, the board again criticized the EPA for its classification of perc and its use of the NCI study results to compute a precise quantitative risk estimate. All but one member of the advisory board felt that perc belonged in IARC category 3, but the board indicated that it was unable to provide an authoritative scientific opinion on perc's carcinogenicity. The board also criticized the late release of the modification of classification, and one member noted that "it [was] rather easy to get the perception that maybe the agency was posturing itself to, in fact, produce an answer that it already thought necessary to have" (Drye, 1991: 106).

In its formal response to the EPA, the SAB suggested a number of additional changes to the HAD, but it indicated that, conditional on these changes, the document would be adequate scientifically as an EPA resource document. Finally, in 1985, the EPA released the final HAD, which indicated that according to a literal interpretation of the IARC criteria, perc would be classified in IARC category 3 (EPA, 1985a). Since the EPA was in the process of finalizing its own weight-of-evidence classification scheme for use in future carcinogen assessments, the HAD also placed perc in the EPA weight-of-evidence category C, "possible human carcinogen." These categories corresponded to the ones in which the HSIA and dry cleaning trade associations asserted perc should be placed.

Shortly after the 1985 release of the final health assessment document, the National Toxicology Program's bioassay results were released, and they caused quite a stir. The NTP inhalation results repeated the NCI mouse gavage liver tumors, and the addition of rat leukemia and kidney tumors changed the weight of evidence by nominally fulfilling the criteria for "sufficient" animal evidence according to the EPA's then pending cancer guidelines. The EPA began preparation of an addendum to the HAD and circulated an interim internal memo that elevated the overall evidence for carcinogenicity to category B2 "probable human carcinogen" (Anderson, 1985). In 1986 the EPA released for public comment the draft addendum to the HAD, which classified perc in category B2 (EPA, 1986a).

The bioassay results and classification were controversial. In essence, the controversy was (and is) over 1) whether each of these responses was genuine or artifactual; 2) whether the responses collectively could support a conclusion that transcends the weaknesses of the individual end points; and 3) whether evidence suggests that some or all of the animal responses are not predictive of possible human responses. Although the reality of the mouse liver tumor response was unquestioned, there were questions about the statistical and biological significance of the rat kidney tumors, and about whether the finding of leukemias was influenced by mortality patterns in the rats.

Again, the SAB review became the forum for addressing the controversy. During its public meeting on the draft, the board was unable to reach consensus on the classification issue or on its interpretation of the rat tumors. However, the board's chairman, John Doull, acknowledged that strict interpretation of the classification criteria led to category B2. Yet the letter that the advisory board ultimately wrote to the EPA administrator to indicate its findings on the HAD came as a surprise to many people who had attended the meeting. In the letter, the SAB disagreed with the interpretation of the rat findings in the HAD and classified perc as a group C possible human carcinogen. John Doull later said, "We were asked to look at the science, but we had a 'semantic problem.' . . . The subcommittee clearly felt that the category should be 'possible' and had to downplay the evidence to that effect" (Drye, 1991: 113). The EPA chose to respond to the SAB by defending its position and refuting the SAB's position on the rat tumors, and to maintain its B2 classification (EPA, 1987).

The controversy over classification then took a new turn as more than two thousand dry cleaners participated in a mail campaign asking their congressional representatives to encourage the EPA to follow the SAB's advice and classify perc in category C. These letters, some accompanied by sympathetic letters from key members of Congress, were sent to EPA administrator Lee Thomas, and industry representatives scheduled a meeting with Thomas to express their concerns about the implications of the B2 classification. The industry's concerns included anticipated regulatory actions (federal and state) tied to classification and increased liability of perc users and manufacturers.

Thomas responded by again asking the SAB for advice, and he indicated his concern about the overriding importance given to classification by regulators, stating, "EPA's classification of a compound has major ramifications beyond its use in EPA's own decision making process. Rightly or wrongly, state environmental decisions and public perceptions of risk are often triggered by an EPA determination to classify a compound as a B2 carcinogen. This black-white interpretation of the classification system is troubling" (Thomas, 1987).

The SAB responded to Thomas by stating:

> The issues regarding the application of the risk assessment guidelines appear not to represent disagreement among scientists about scientific evidence but, rather, the consequence of attempting to fit the weights of evidence into necessarily arbitrary categories of risk. Since the weights of evidence, and uncertainties associated with such evidence, for perchloroethylene and other compounds fall within a range of scientifically defensible choices, it may not be possible, in some instances, to fit them neatly into only one risk category. Moreover, the

> more incomplete the data, the less precision one can expect in classifying a compound within EPA's cancer guidelines. In addition, the type of evidence that places a compound in a particular category may vary considerably from substance to substance within that category. For perchloroethylene, as with trichloroethylene, the Science Advisory Board concludes that the overall weight of evidence lies somewhere on the continuum between the categories B2 and C of EPA's risk assessment guidelines for cancer (Doull, 1988).

The board also indicated that it shared Thomas's concern about the "black-white" regulatory implications of B2 versus C classification and recommended that the EPA reevaluate the need to change its classification system and methods of characterizing uncertainty, stating: "The distinction between the B2 and C categories can be an arbitrary distinction on a continuum of weight-of-evidence. . . . From a scientific point of view, it seems inappropriate for EPA and other agencies to regulate substances that are classified as B2 and not to consider regulations of compounds classified as C, regardless of the level of human exposure. . . . A substance classified as C (limited evidence in animals) for which human exposure is high may represent a much greater threat to human health" (Doull, 1988).

The EPA did not revise or finalize the draft addendum. Nonetheless, the EPA continued to propose regulations for perc between 1989 and 1991 on the basis of a B2 classification (for example, in establishing Resource Conservation and Recovery Act [RCRA] land disposal requirements and treatment standards, Comprehensive Environmental Response, Compensation and Liability Act [CERCLA] reportable quantities, and drinking water standards). In their public comments on these regulations, the HSIA and dry cleaning industry trade associations continued to pressure the EPA into classifying perc in group C. In 1991 the EPA addressed the public comments by drafting a "Response to Issues" document, summarizing its view of updated evidence about perc's carcinogenicity for consideration by the SAB (EPA, 1991c). This document reviewed new metabolic evidence that provided explanations for the tumors observed in the NCI and NTP animal studies. In this document, the EPA concluded that "although the relevance of some of the data is less than certain, the inclusive animal data for perchloroethylene taken as a whole, along with the considerations of inadequate human data, information on metabolism, and mutagenicity data on metabolites, perc can most logically be categorized as a Group B2 probable human carcinogen" (EPA, 1991c: 59):

The SAB reviewed this draft and disagreed. In its response to the EPA administrator, the board states:

> We do not consider the evidence strong enough to classify this compound as a probable human carcinogen (i.e., B2); on the other hand, the evidence for carcinogenicity *is stronger* than for most other compounds classified as possible human carcinogens (i.e., C). Therefore, in the spirit of the flexibility encouraged by the Guidelines, our best judgment places this compound on a continuum between these two categories. . . . [Perc] is an example of a chemical for which there is no compelling evidence of human cancer risk, but for which reductions in unnecessary human exposure might well be prudent. The available scientific information does not suggest to us the same regulatory responses that would be appropriate for a chemical whose bioassay responses were clearly relevant to human cancer (Loehr, 1991).

Recently, the IARC reconsidered its classification of perc in preparation for upgrading its *Monographs on the Evaluation of Carcinogenic Risks to Humans.* Based on its review of several epidemiological studies (primarily those discussed in Weiss [1995a]), the IARC found that there was limited evidence in humans. This limited evidence was considered to be just over the line, but it led to the reclassification of perc into group 2A, "probably carcinogenic to humans." At the same time, the IARC adopted a 2B, "possibly carcinogenic to humans," classification for dry cleaning. This apparent inconsistency has led one observer of the process to point out a potential deficiency in the IARC classification system (Weiss, 1995b).

The EPA is currently in the process of revising its carcinogen classification scheme, at least in part due to compounds like perc that have proven difficult to classify under the existing scheme. It is unclear whether or not there will ever be definitive evidence regarding perc's ability to cause cancer in humans. The recent animal evidence provides support for hypotheses about mechanisms of perc's tumorigenesis, but it is still equivocal as to the relevancy of these data to humans. When the hypothesized mechanism is of uncertain relevance to humans, society faces the long-standing issues of whether it is enough to see a carcinogenic response in animals to consider a substance to be a potential human carcinogen, and how much caution is appropriate in risk management (that is, when to apply the "precautionary principle").

| Perc as a Precursor to Smog

Throughout the controversy regarding perc's classification, EPA and the Occupational Safety and Health Administration (OSHA) proposed and promulgated a number of regulations related to releases of perc from dry cleaners. In some cases, dry cleaners were regulated due to a qualitative assessment of perc's hazard. One example of this was the regulation of perc as a volatile organic compound. Two of the major precursors of photochemical smog (tropospheric ozone) formation are nitrogen oxides and VOCs. The amounts of ozone formed depend on a complex set of chemical reactions that occur in the presence of sunlight. Tropospheric ozone is responsible for the smog sometimes observed in urban areas, and it is considered to be a respiratory hazard.

In 1977 the EPA released a policy on the control of VOCs. Under the policy, all VOCs were considered to contribute to tropospheric ozone formation unless they were specifically exempted due to their negligible photoreactivity. The VOCs policy explicitly stated that a number of compounds that had only negligible photochemical reactivity, including perc, would not be exempted, because they had been identified or implicated as being carcinogenic, mutagenic, or teratogenic. In the case of perc, the policy stated that it had "been implicated as being a carcinogen" (EPA, 1977: 35315). The policy recommended against excluding perc from control asserting that "it would be unwise to encourage [its] uncontrolled release" (EPA, 1977: 35315).

As required by the VOCs policy, states began including perc as a VOC in their state implementation plans aimed at reducing smog, and many states required dry cleaners to reduce their emissions of perc. In 1978 the EPA published a control technique guideline (CTG) document on perc dry cleaners to assist states in controlling emissions from existing sources and to recommend carbon adsorbers as a reasonably available control technology (RACT) (EPA, 1978). (Carbon adsorbers are add-on control devices that strip perc out of the air normally exhausted from the dry cleaning machine and then recycle most of the recovered perc back to the system.)

The industry responded to the VOC regulations largely by reducing its emissions of perc. Compliance was not difficult, because the regulation occurred when the industry was already reducing its consumption of perc (and consequently its emissions of perc) in response to cost-saving opportunities (associated with more efficient dry-to-dry machines and carbon adsorption technology) and declining consumer demand for dry cleaning services. The ability to decrease perc releases with more efficient machines and to recover and recycle perc using carbon adsorbers meant a substantial cost savings on solvent purchases. The industry attributed the decline in consumer demand, which started in the mid 1970s, to the increasing popularity of wash-and-wear fabrics made with synthetic fibers. A trade publication, the *Chemical Marketing Reporter,* characterized domestic dry cleaning as "on the decline" in 1976 (CMR, 1976), and then indicated in 1979 that consumer demand for dry cleaning was expected to increase as fabrics made with natural fibers (such as wool blends and silks) regained popularity (CMR, 1979). Dry cleaning demand decreased between 1974 and 1981 but started increasing substantially starting in 1982 (EPA, 1991b). Seitz (1994) attributes the increased popularity to consumer dissatisfaction with synthetic fabrics and a dramatic increase in the number of women in the workforce.

In 1983 the EPA issued a notice to propose that perc be exempted from the list of contributors to tropospheric ozone (EPA, 1983). The proposal was based on the results of smog chamber studies that showed perc was less reactive than a substance (ethane) that had been exempted in the 1977 VOCs policy. The smog chamber is a controlled environment that allows scientists to estimate the lifetime of a substance and rate constants for degradation reactions. The chamber studies suggested that perc degrades slowly. It is one of the few compounds that degrades in the upper troposphere, which makes it both an insignificant tropospheric ozone air pollution contributor and a negligible stratospheric ozone depleter (Wang, Blake, and Rowland, 1995). The dry cleaning industry did not pressure the EPA to finalize the 1983 proposal, because at that point control devices had already been installed (Risotto, 1994).

With the 1990 amendments to the Clean Air Act, additional pressures were placed on states to control VOCs, and this led to some concern from dry cleaners that additional emissions reductions would be required of them by states (Spaw, 1990). In addition, the 1978 CTG recommending carbon adsorption technology as a reasonably available control technology had not been updated to reflect the technological changes that had occurred in the industry. In particular, new "state of the art" dry cleaning machines contained built-in refrigerated condensers that the EPA now believes to be more efficient for controlling vented emissions in actual practice than carbon adsorbers (EPA, 1993). Since perc's negligible photochemical reactivity had been acknowledged in the 1983 proposal, the HSIA (with support from dry cleaning trade associations) petitioned the EPA in 1992 to exempt perc as a VOC, and the EPA responded to the petition by proposing the exemption (EPA, 1992a). In this proposal, the EPA indicated that the 1983 proposal had not been finalized due to comments that raised concerns about perc as a toxic air pollutant (EPA, 1992a).

The EPA finalized its 1992 proposal in February 1996 (EPA, 1996b). The delay in finalizing the proposal appears to have been related to comments the agency received from environmental groups like the Natural Resources Defense Council (NRDC). The NRDC sought delay or withdrawal of the 1992 proposal based on concerns about the toxic effects of perc that it raised in comments on the 1983 proposal (Sheiman, 1992). In particular, environmentalists felt the proposal should not be issued until the EPA had established that the public health is adequately protected by controls on emissions from all sources (for example, dry cleaners, degreasers, and sources that use perc in paints, adhesives, pharmaceuticals, printing inks, and dielectric fluids).

The issue of exemption posed a dilemma for the EPA. On the one hand, the EPA was concerned about increased exposure to perc (which it considered to be carcinogenic) from sources that might switch from nonexempt VOCs to perc and new sources that might start using perc following exemption. On the other hand, regulating perc as a contributor to tropospheric ozone meant that emission credits obtained by controlling perc could be used to offset increased emissions of chemicals that actually do increase levels of tropospheric ozone (see Lake [1992] for an example of this in San Diego County), and possibly for chemicals that the EPA considers to be at least as toxic as perc. Thus, not exempting perc created an implicit and uncharacterized risk-risk trade-off.

Regardless of controversy surrounding perc's exemption, the regulation of perc as a contributor to tropospheric ozone helped to motivate some dry cleaners (particularly those in areas that had failed to meet national air standards for ozone) to adopt more efficient machines, to install emissions control devices, and to reduce their consumption of perc.

| Regulating Perc as a Hazardous Air Pollutant

In 1980, following the regulation of perc as a VOC, the EPA proposed a new source performance standard (NSPS) for perc dry cleaners, "based on [perc's] role as an ozone precursor" (EPA, 1980: 78175). This proposal required that new professional dry cleaning facilities install the reasonably available control technology (carbon adsorbers) and indicated that the EPA was at that time investigating the possibility of listing perc as a "Hazardous Air Pollutant" (HAP) under Section 112 of the Clean Air Act and establishing a national emission standard.

While EPA may have been developing the emission standard in the early 1980s, the agency began formal regulatory activity following release of a 1985 memo that indicated the Carcinogen Assessment Group's intention to classify perc as a B2 carcinogen (Anderson, 1985). At that time, the EPA published a notice of intent to list perc as a hazardous air pollutant and presented the results of its preliminary quantitative risk assessment for perc (EPA, 1985b). The EPA estimated an annual cancer incidence from exposure to perc emitted to ambient air by dry cleaners of 3.2 cases per year nationwide, more than half of the 5.3 cases per year estimated for exposure to perc from all sources (EPA, 1985b). It also provided an estimate of one in ten thousand (1.5×10^{-4}) as the maximum individual risk to nearby residents (defined as the "additional risk of cancer to an individual continuously exposed to the highest modeled ambient concentration for a 70-year lifespan" [EPA, 1985b: 52882]).

The notice indicated that the EPA would decide whether to add perc to the list of HAPs "only after studying possible techniques that might be used to control emissions of perc and after further improving the assessment of the public health risks" (EPA, 1985b: 52883). The notice, which came out while the Carcinogen Assessment Group was preparing an addendum to the health assessment document, only considered the listing of perc and did not propose any control requirements. Consequently, it is not surprising that the industry did not spend a great deal of time contesting the quantitative risk assessment. The industry focused instead on its position that the available evidence was too weak to support perc's human carcinogenicity.

The EPA continued development of a national emission standard and internal memos estimated that the candidate proposal, requiring carbon adsorbers, would cost an average of $13 million per life saved: approximately $8 million per life saved for commercial cleaners, a net savings for large industrial cleaners, and approximately $45 million per life saved for coin-operated cleaners (EPA, 1986c). These estimates were never included in a proposed rule, however, because the EPA did not take final action to list perc as a HAP, possibly because it was waiting for final approval of the health assessment document. In 1988 a draft cost-benefit analysis for the regulation of dry cleaners found that establishing an emission standard for perc dry cleaners would be beneficial, largely based on the decreased occupational risks, and that from a federal perspective the need to establish a standard appeared to be secondary to the need to regulate occupational exposures (EPA, 1988).

In 1990 Congress ultimately settled the unresolved issue of perc's status as a hazardous air pollutant under Section 112 of the Clean Air Act by listing it in the revised statute along with 188 other substances. There was a general sense that regulation of toxic air pollutants was proceeding too slowly, and consequently Congress designed a solution by listing the HAPs, requiring the application of best available technology as a first pass, and turning to residual risk later. Perc made it on to the list of HAPs because it was on the Superfund Amendments and Reauthorization Act Section 313 and CERCLA Section 104 (reportable quantities) lists, and ultimately its listing traces back to its 1976 designation as a priority pollutant (Voorhees, 1989). (The priority pollutants were established under a consent decree requiring the EPA to study and develop regulations under the Federal Water Pollution Control Act. How it was selected as a priority pollutant is undocumented.) At approximately the same time, a private citizens group from Oregon sued the EPA because it had failed to issue a final rule on the 1980 proposed new source performance standard for perc dry cleaners. The lawsuit was settled with the establishment of a consent decree that required the EPA administrator to establish a national emission standard for perc dry cleaners within two years of the enactment of the 1990 amendments to the Clean Air Act.

In December 1991 the EPA simultaneously proposed a national emission standard and withdrew the new source performance standard that it had proposed in 1980 (EPA, 1991d). In its proposed rule, the EPA did not present the results of a quantitative risk assessment, because the emission standard was technology based. Instead, hazard and exposure are discussed qualitatively in the rule which concluded that perc emissions from commercial dry cleaners presented a "threat of adverse effects" and should be controlled (EPA, 1991d: 64389). The analyses used to support the regulation (EPA, 1991a, b) examined the cost-effectiveness of the rule in terms of dollars per kg of perc recovered, but not in terms of risk. Overall, the dry cleaning trade associations cooperated with the EPA in its development of the emission standard and the promulgation of the final regulation (EPA, 1993). The control requirements established by the standard will be fully in effect in 1997.

| Protection of Workers

In 1971 the Occupational Safety and Health Administration established the first federal regulatory standard based on concern about adverse health effects associated with exposure to perc. As required under Section 6(a) of the Occupational Safety and Health Act, OSHA promulgated permissible exposure limits (PELs) for perc and over a hundred other substances, without performing a risk assessment or allowing public hearing or comment. PELs indicate the maximum permitted level of worker exposure averaged over 8 hours or 15 minutes, as well as ceilings that should never be exceeded. OSHA established an 8-hour PEL for perc of 100 ppm to protect workers from liver damage, which was believed to be the most sensitive biological effect at the time.

In 1988 OSHA proposed to lower its general industry PEL for perc to 50 ppm in a "generic rulemaking" based on possible central nervous system effects (OSHA, 1988). Following comments on its proposal, however, OSHA further lowered the PEL to 25 ppm for perc in its final rule, based on concerns about carcinogenicity (OSHA, 1989). OSHA apparently did not conduct its own risk assessment for perc, but the results of an external risk assessment conducted by Dale Hattis for the National Institute of Environmental Health Sciences figures prominently in the rule. The rule states "The risk assessment conducted by Hattis [Hattis et al., 1986] estimates that there is an excess lifetime cancer mortality risk of 45 deaths per 1,000 workers exposed for 45 years to the current 100-ppm [time weighted average] PEL. Clearly, this high risk of mortality represents a significant risk. At the proposed level of 50 ppm, Dr. Hattis estimated the excess lifetime risk to be 27 deaths per 1,000 workers. OSHA concludes that this assessment and the underlying evidence clearly indicate that a further reduction in the PEL is necessary" (OSHA, 1989: 2688). A number of plaintiffs, including a dry cleaning trade association, the International Fabricare Institute (IFI), and the HSIA, sued OSHA, asserting that it did not meet procedural guidelines in establishing the lower PELs.

In July 1992 a panel of three judges of the U.S. Court of Appeals for the Eleventh Circuit agreed with the plaintiffs, and it vacated the lower PELs (*AFL-CIO v. OSHA,* 965 F.2d, p. 962, July 7, 1992). In its opinion, the court found that "OSHA's analysis of [perc] is a prime example of the problems with OSHA's approach to this rulemaking." Industry groups argued that OSHA's determination that perc presents a significant cancer risk was not supported by substantial evidence, while unions argued that the final rule level still left workers exposed to a significant cancer risk. The record left the reviewing court unable to decide. OSHA's request to have the full Eleventh Circuit Court rehear the case was denied, and the case was not appealed to the Supreme Court (DOL, 1993). Consequently the PEL for perc returned to 100 ppm, except in those states that had adopted the lower PEL by state statute. In addition, OSHA withdrew the PELs it had similarly proposed (OSHA, 1992) for construction, maritime, and agricultural employees using a general industry rulemaking. Although OSHA has not proposed to lower the PEL for perc to date, it is likely to reconsider perc's permissible exposure limit soon.

| Perc as a Hazardous Waste

Dry cleaners became subject to hazardous waste regulations implemented by the EPA with the establishment of standards for small quantity generators under the Resource Conservation and Recovery

Act in 1986 and the establishment of reportable quantity rules for the Comprehensive Environmental Response, Compensation, and Liability Act (EPA, 1989).

Under RCRA the EPA regulates the accumulation, transportation, long-term storage, disposal, and management of hazardous waste, but RCRA does not cover the on-site recycling that most dry cleaners perform. Prior to the establishment of regulations for facilities generating between 100 and 1000 kg per month of waste, most dry cleaners disposed of their solid waste legally in municipal landfills.

In 1986 the EPA proposed and finalized land disposal restrictions for perc (EPA, 1986d). At that point, dry cleaners had three options for dealing with their perc solid waste: 1) to dispose of waste in permitted hazardous waste landfills, 2) to ship waste to be incinerated, or 3) to contract with an approved recycler who would process the waste to recycle the perc. In 1986 the IFI estimated the costs of these options for an average dry cleaner would be: for option 1, $5,000 a year until November 1988 and higher thereafter when the RCRA land disposal bans came into effect; for option 2, $8,000 to $13,000 per year; and for option 3, $800 to $1,200 per year (Wolf and Myers, 1987; Meijer, 1994). (The final rule prohibited the land disposal of waste containing more than 0.079 ppm of perc, or 0.0000079 percent, as of November 1988, although it allowed land disposal of waste containing up to 10,000 ppm of perc, or 1 percent, until that date [EPA, 1986e]).

By far, the least costly option was to contract with a recycler. However, in 1986 two dry cleaning trade associations, the IFI and the Institute of Industrial Launderers, were concerned with the lack of recycling services in some areas of the U.S., and they petitioned the EPA to exempt filters from regulation (Meijer, 1994). Exempting the filters would have allowed many dry cleaners to continue to dispose of filters in municipal landfills. The EPA never formally acted on the petition, and it was ultimately withdrawn by the dry cleaning associations when it was found that recyclers were able to provide service to almost every interested dry cleaner in the country (Meijer, 1994). Currently it is estimated that recyclers handle the majority of the dry cleaning solid waste generated (CEC, 1992). Overall, the widespread adoption of recycling brought dry cleaners into compliance with the RCRA quickly and dramatically reduced the amount of perc going into landfills.

Until recently, the effect of the Comprehensive Environmental Response, Compensation, and Liability Act on the dry cleaning industry was most easily characterized by looking at individual facilities. For some facilities, the liability created under CERCLA has been devastating. When dry cleaning facilities or former facilities become classified as hazardous waste sites (as they have been in a number of states), the owners of these facilities become liable under CERCLA and RCRA for very costly cleanups. In addition, private parties are requiring cleanups at existing dry cleaners in conjunction with sales, refinancings, and commercial operation. Risk assessments have been prepared for many individual facilities. Since the sites are typically small, they do not generally qualify for federal cleanup money and many dry cleaners are put out of business. This has become a major concern for the industry, since up to 90 percent of dry cleaning facilities could involve some level of site contamination (Meijer, 1994). Some dry cleaners are installing safety troughs, diking systems, and floor sealants to prevent future contamination. However, while these control devices provide insurance against future liability, they do not save individual dry cleaners from retroactive liability.

The threat of liability appears to have motivated the industry overall to take a more proactive stance toward protecting the environment. While the dry cleaning industry still holds that there is insufficient evidence to classify perc a human carcinogen, industry leaders urge individual dry cleaners to minimize their releases of perc into the environment.

A major waste issue for dry cleaners stems from the formerly legal practice of discharging perc-contaminated waste water into city sewer systems. In 1990 dry cleaners were the suspected sources of perc contamination in several city wells in Merced, California. In July 1990 the IFI presented its recommendations for disposing of contaminated waste water, in response to "an increasing number of reports of ground water and soil contamination" (IFI, 1990). The IFI indicated that alternative methods existed for recycling perc from the waste water, and that these posed fewer environmental risks than discharging to a sewer or septic system. The IFI (1990) identified the use of an evaporator or an aerator (to volatilize the perc from the water) as the most environmentally sound option for disposing of perc-contaminated waste water. The main problem with using these devices at the time was that the waste water was classified as a hazardous waste under RCRA, and consequently using an evaporator was considered "treating the waste," which required a facility to go through an extensive and expensive permit process. After two years of going back and forth between the dry cleaners, the states, manufacturers of dry cleaning equipment, and various EPA offices, the EPA decided to allow the treatment of separator water without requiring RCRA permits (Lowrance, 1993). This appears to be a case in which the risk-risk trade-off (water versus air) was informally evaluated, and the decision facilitated a reduction in perc discharges to water.

The most threatening aspect of hazardous waste regulation for dry cleaners is the potential for private party actions. The dry cleaning trade associations are currently trying to persuade Congress to establish a separate cleanup fund in CERCLA's reauthorization that would be used specifically for remediation of dry cleaning sites. Under the most recently proposed plan, to generate a cleanup fund dry cleaners would pay a fee of $7.95 per gallon of perc purchased, plus a per site fee that would be set according to the number of employees, and they would adopt a number of pollution prevention practices. In return, dry cleaners are asking to be released from retroactive liability under federal and state RCRA and CERCLA actions (IFI, 1995). The proposed pollution prevention practices include a complete phaseout of all transfer machines within five years, and requirements that all dry cleaners operate dry-to-dry machines with integral refrigerated condensers, install dikes, seal floor surfaces, use closed and direct-coupled delivery systems to minimize spills during transfers, and treat *all* perc-containing waste (including waste water), no matter what the quantity, as hazardous waste. Florida passed its own cleanup legislation that places a $5 per gallon tax on perc and takes 1.5 percent of gross receipts to establish a statewide cleanup fund. Several other states have adopted or are considering similar approaches.

| Indoor Air and Pollution Prevention

In 1986, following release of the health assessment document, the Office of Toxic Substances (OTS) performed a risk assessment that combined the unit risk estimated in the 1986 HAD with perc population exposure data it had developed that were not specific to dry cleaning.[2] The assessment reported point estimates of the upper-bound individual risk ranging from 1×10^{-4} to 1×10^{-8}, and the upper-bound estimate of additional cancer cases was on the order of 600 cases per year nationwide (Rhomberg, 1986). This assessment included a lengthy discussion about the uncertainties in the estimates, and concluded that "it is difficult to gauge the magnitude of these uncertainties or to provide any reasonable range over which risk might vary" (Rhomberg, 1986:19). This risk assessment apparently was not used to support any regulatory activity.

Between 1988 and 1992, the OTS considered the issue of residual levels of perc in dry cleaned clothes. These efforts focused on characterizing exposure and identifying techniques for reducing levels of perc introduced into the home (Tichenor et al., 1990) as opposed to predicting risks. This study (and another similar one reported by Thomas et al., 1991) measured elevated levels of perc indoors following the introduction of freshly dry cleaned clothes into test homes. However, the levels were relatively low (implying relatively low risk), and efforts to identify means of reducing perc residuals (either by the dry cleaner or the consumer) have been unsuccessful (EPA, 1992b). Thus it is not surprising that this issue has not been the basis for regulatory activity or the subject of a major risk assessment.

In 1992 the EPA's Office of Pollution Prevention and Toxics (OPPT) (which succeeded the OTS) initiated a voluntary cooperative program, called Design for the Environment (DfE), to promote the incorporation of environmental considerations and risk reduction ideas into industrial processes, products, and technologies. One major component of DfE projects is the development of a cleaner technology substitutes assessment (CTSA) intended to help an industry evaluate pollution prevention options by comparing the risks, exposures, performance, costs, pollution prevention opportunities, and energy conservation associated with different solvents and technologies.

Since the program is voluntary, the first step involves identifying interested parties and forming a project team. The EPA kicked off the DfE dry cleaning project in May 1992 when it sponsored a roundtable discussion with industry representatives and other interested parties to discuss pollution prevention in the industry (EPA, 1992b). Two impor-

[2] This section is largely based on Jehassi (1995) and Risotto (1995).

tant events occurred prior to the roundtable. First, the New York State Department of Health released a report about levels of perc measured in six apartments above dry cleaning establishments and brought attention to the possible risks to residents (NYDOH, 1991). Second, the California Regional Water Quality Board released its study about perc ground water contamination (EPA, 1992b). These issues were discussed at the roundtable meeting, which was exploratory and somewhat adversarial.

Shortly after the roundtable, the EPA received results of a study sponsored by Greenpeace demonstrating that a new technology called multiprocess wet cleaning offered a potentially viable alternative to perc dry cleaning. This technology cleans "dry clean only" garments using water as the solvent in a relatively labor-intensive process that involves hand washing of the clothes and may or may not use mechanical drying. The EPA's DfE team members designed a series of demonstrations of the multiprocess wet cleaning technology to evaluate its feasibility as an alternative to perc, and they coordinated a demonstration of this technology in New York at the Neighborhood Cleaners' Association in November and December 1992. The results of the demonstration appeared promising for at least some percentage of garments, and led to the initiation of additional research on commercial viability, customer acceptance, and cleaning performance.

By early 1993 the dry cleaning industry had committed itself to participation in the DfE project. Several other interested parties also joined the project, including environmental groups and labor unions (EPA, 1995). Once the group was established, the consideration of different alternative technologies began internally at EPA. In addition to multiprocess wet cleaning, other alternatives that have been or are being considered to some degree include:

- Machine wet cleaning of clothes in a machine that simulates hand washing and uses several substances to prevent fabrics from shrinking and fading. The industry has actively participated in and helped to fund research on this technology.
- A new machine that uses liquid CO_2 as the cleaning solvent, demonstrated for the first time at a national dry cleaning trade show in 1995. Efforts to make this relatively expensive technology cost competitive are underway. The research for this

technology has been sponsored by the Department of Energy and not the DfE parties.

- Improved (less flammable) petroleum solvents that may compete favorably with perc, based on their lower cost. Although these solvents contribute to tropospheric ozone and are subject to VOC regulations, they may not have the same carcinogen stigma as perc (because they have not been tested) and dry cleaners may consequently look on them favorably as substitutes. The explosive nature of these solvents remains an issue as demonstrated by the fire that occurred in March 1995 at a German dry cleaning business (Wentz, 1995).
- Three hydrochlorofluorocarbons are possible substitutes for CFC-113 and TCA: HCFC-141b, HCFC-123, and HCFC-225. They do not, however, appear to be good substitutes for perc because they ultimately will be phased out under the Montreal Protocol and amendments, due to their ability to deplete stratospheric ozone and because their use poses other risks (Wolf, 1992). For example, HCFC-141b is flammable at concentrations required for dry cleaning, initial toxicity testing results suggest that HCFC-123 may cause pancreatic and testicular tumors in male rats, and HCFC-225 may also be toxic, although it has been approved as a substitute for CFCs under the EPA's Significant New Alternative Policy Program (EPA, 1994b). Hydrofluorocarbons and fluorocarbons offer other possibilities, but like the HCFCs they are unlikely to be used for dry cleaning.

A series of performance demonstrations is under way to provide more cost and performance data for the two wet cleaning processes and for a more efficient drying process that uses microwaves to evaporate water from the clothes. The cleaner technology substitutes assessment is also expected to include analyses of the risks and costs of conventional perc dry cleaning to facilitate comparisons between the different technologies.

In 1992 the DfE project team started the preliminary work on the perc risk assessment for the CTSA and focused on the risks posed by dry cleaners located in residential buildings. This focus arose following criticism from the New York Department of Health that the proposed national emission standard for perc as a hazardous air pollutant did not address indoor air and did not regulate dry cleaners located in residential buildings. The OPPT risk assessment combined the 1986 health assessment document's unit risk with exposure estimates based on measurements made in the New York Department of Health's study of single-day perc measurements in six apartments located over dry cleaners. The results of this screening assessment estimated upper-bound lifetime individual cancer risks as high as 1×10^{-3} for highly exposed individuals, and led to the conclusion that the risks could be high enough to warrant further investigation of actions that could be taken to reduce those exposures and risks (EPA, 1992c).

After several years of work on the next phase, the CTSA is still under development and going through internal review. The delay in releasing the CTSA is a result of resistance on the part of industry stakeholders who believe that quantitative risk estimates are not required for the objectives of the DfE program. Industry leaders argue that the goal of the CTSA is to compare technologies and that numerical risk assessments might be harmful to the industry and counter-productive in the pollution prevention process. In a letter to Elizabeth Parker, then director of the DfE program, the HSIA asserted that:

> If OPPT has not undertaken the in-depth assessment for [perc] recommended in the [National Academy of Science] report or anticipated in EPA's new guidelines, but is considering "default" assumptions with the linearized multi-stage model, then it is important that the general public is made aware of the uncertain character of the risk estimates, and that the true risk may be anywhere from zero to the statistically contrived "upper bound." Certain members of the DfE "stakeholder" group might be expected to use any risk estimates that may be developed in vigorous public campaigns without defining the uncertainties in the values. This public display of supposed risks, especially if based on inappropriately simplistic assessment procedures, could harm the DfE program, embarrass EPA, and unfairly harm the dry cleaning industry which is engaged in significant efforts to reduce [perc] releases to the environment. (Voytek, 1994)

The industry concern was intense enough to motivate four members of Congress to urge EPA administrator Carol Browner to use "good science" in the CTSA and to withhold its release if it contains "flawed numbers" that could harm the dry cleaning industry (Portman et al., 1995). In contrast, environmentalists argue that the risk assessment results should be released quickly so that the public can be informed about what they consider to be a "severe, immediate threat to public health and the global environment" (Rice, 1994).

Thus, this is a case in which environmentalists appear to be advocating the release of quantitative risk assessment results, while industry groups oppose it. The industry argues its opposition is not about the use of risk assessment per se, but instead is based on its concern that the results do not utilize all of the biological data available and are based on very limited exposure measurements. Although the DfE program is ongoing and the CTSA is yet to be released, it appears that the threatened release of a risk assessment has been very important in keeping the industry engaged and participating in the process. The issue of characterizing the risks has been central to the DfE group's discussions, and ultimately it may be more important than the CTSA if and when it is released.

Overall the program has been successful in promoting pollution prevention in the industry. In November 1995 environmental and labor groups reached an agreement with dry cleaning industry leaders to form the Professional Wet Cleaning Partnership (PWCP). In spite of the widely different views held by the PWCP partners, all acknowledge that concerns about perc's potential carcinogenicity are driving the industry to change, and all agree that it is desirable to increase familiarity and expertise with professional wet cleaning (PWCP, 1995). One likely outcome of this effort will be a shift of some perc dry cleaning to wet cleaning, although the degree of the shift is unknown.

One key issue yet to be resolved is what percentage of "dry clean only" clothes can be wet cleaned. Current estimates range from 40 percent to 70 percent (Wilkin et al., 1995), and environmental groups believe the goal should be 100 percent. The main "problem fabrics" include wools, acetates, and some clothes with water-soluble dyes. Recognizing that wet cleaning is possible at least some of the time but that it may be prohibited by garment labels that read "dry clean only," members of the PWCP collectively asked the Federal Trade Commission to consider reopening the rule on garment care labeling requirements (EPA, 1995).

| Pollution Reduction by the Industry

The amount of perc used by the dry cleaning industry is decreasing. Figure 4.2 shows estimates of the historical levels of U.S. market demand for perc and the corresponding dry cleaning demand for perc since 1960 (ITC,1960-1993; CEC,1992; SRI, 1994). Market demand is defined as U.S. production plus imports minus exports, and there are no known natural sources of perc. The points labeled CEC provide estimates of amounts of perc sold to the dry cleaning industry, and those from SRI provide estimates of perc use by the dry cleaning and textile processing industries combined (textile processing demand is relatively small compared with the demand from dry cleaning). The perc used for dry cleaning has historically accounted for at least 50 percent of perc production. Perc is or has been used in the aerosol industry, for metal cleaning and degreasing, and as an intermediate in the production of chlorofluorocarbons (CFCs), hydrofluorocarbons (HFCs) and hydrochlorofluorocarbons (HCFCs).

Since the 1970s the industry has cut its consumption of perc in half (Figure 4.2). While demand for dry cleaning services has varied

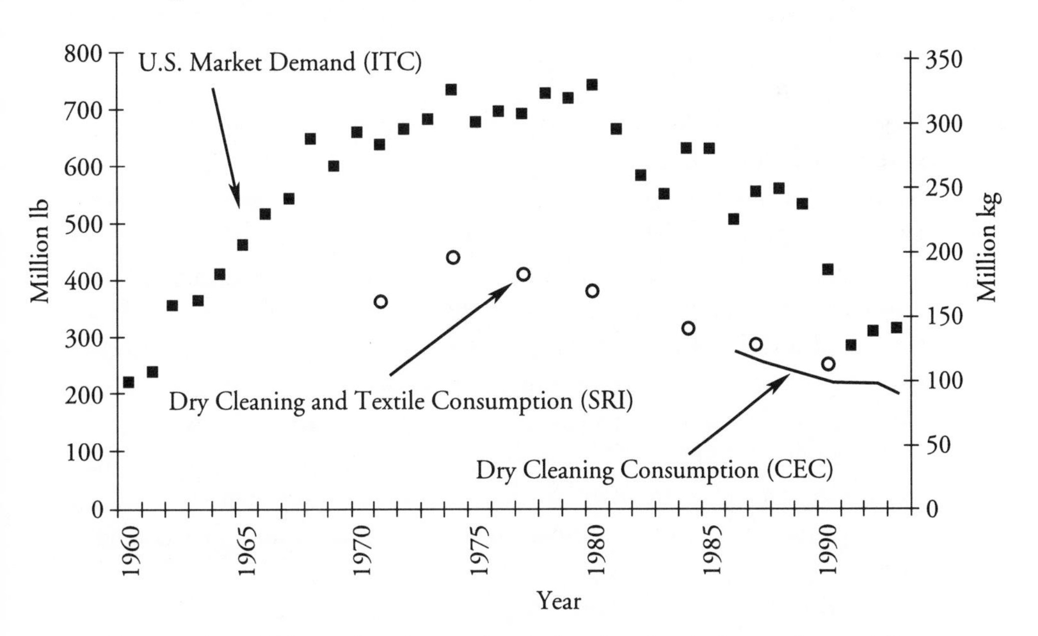

FIGURE 4.2 | U.S. market demand and dry cleaning industry demand for perchloroethylene.

during this period, it is currently estimated to be at least five percent higher than it was in the early 1970s (EPA, 1991a). Thus, dry cleaners are using less perc to clean more clothes now than they did twenty years ago, and this increased efficiency translates into less pollution. This trend may be expected to continue as more and more individual dry cleaners replace their existing machines with more efficient ones. For example, fugitive emissions associated with transferring clothes will be eliminated as dry cleaners replace transfer machines with dry-to-dry machines. Emissions associated with venting perc to the atmosphere will be reduced as dry cleaners purchase control devices or nonvented machines. Emissions from loading the machines will be reduced by adding enclosures or control technology to the machine doors. Emissions from spills will be reduced with the installation of safety troughs and floor sealants and with the use of direct-couple delivery systems. Finally, emissions of perc into water systems will diminish as facilities install evaporators or treat waste water as hazardous waste. As the industry evaluates potentially promising new technologies like wet cleaning and liquid CO_2, perc use will continue to decline as well. This slow shifting of solvents is consistent with the history of the industry.

| The Role of Risk Assessment

Concerns about risks associated with dry cleaning have clearly influenced the solvent and technology choices made by the industry. The industry has responded to risk concerns by 1) switching solvents (for example, by shifting from petroleum solvents to perc), 2) reducing its emissions as required by regulation (for example, by installing carbon adsorbers to comply with the VOC regulations), 3) reducing its emissions voluntarily (installing safety troughs, floor sealants, direct-coupling devices), 4) recycling solvent at the industry level (instead of sending hazardous waste to landfills), 5) resisting attempts by regulatory agencies (CPSC, OSHA, EPA) to classify perc as a carcinogen, and 6) cooperating with regulatory agencies (for example, with the EPA when it established the national emission standard and in the DfE program).

While dry cleaners had obvious economic incentives to avoid excessive use of perc, federal regulations aimed at protecting human health have significantly influenced the industry. One representative of the dry cleaning industry attributed the reduced consumption of perc since 1970 to 1) the regulation of perc as a volatile organic compound, 2) the 1985 EPA proposal to list perc as a hazardous air pollutant based on positive animal carcinogenicity tests, and 3) concern about possible lawsuits from employees and local residents who might later develop cancer (Wolf and Myers, 1987: 22). Another industry publication suggests that "increased efficiency has resulted in large part from controls imposed under state and local smog and 'air toxics' regulations and federal regulation of the disposal of dry cleaning wastes" (CEC, 1992: 41).

Although uncertainty and controversy remain about perc's ability to cause cancer in humans, the possibility that it might do so has clearly motivated a great deal of regulation and industrial response. The EPA continued to label perc as a B2 carcinogen when proposing regulations until perc was listed as a HAP in the 1990 Clean Air Act Amendments. While the industry trade associations did not necessarily object to regulatory control requirements, they continued to pressure the EPA to classify perc in category C. Dry cleaners are very concerned that the public views a B2 classification as a definitive indication that something "is a carcinogen," while C classification does not. Recent research suggests that no public consensus exists regarding the distinction between these classifications and in fact that both lay people and experts have a difficult time interpreting their meaning (see, for example, Spedden and Ryan, 1992).

The EPA's difficulties in classifying perc and the industry's response to its classification demonstrate the importance that has historically been placed on classification. If the EPA had not perceived the need to classify perc in group B2 to support regulation, then regulation of perc under several statutes (particularly the national emission standard for perc as a hazardous air pollutant) might have occurred earlier than it did. However, without classification the controversy also might have merely shifted to another issue, such as the action to list perc as a HAP. Whether designating perc as "hazardous" would have been as vigorously contested as the "carcinogen" label is a matter of speculation.

The EPA's extended regulation of perc as a contributor to tropospheric ozone suggests that the agency perceived its existing "toxics" regulations of perc to be weak. This perception was apparently strong enough for the EPA to justify the implicit risk-risk trade-off made in this case between perc (which is not a contributor to tropospheric ozone) and other substances (which may contribute to tropospheric ozone and may be more or less toxic than perc).

Overall, complete quantitative cancer risk assessments have played a relatively minor role in the environmental regulation of dry cleaners using perc. The lack of a risk assessment undermined OSHA's new PELs but had no apparent effect on the establishment of solid waste regulations. The EPA's preliminary analysis for listing and regulating perc as a hazardous air pollutant provided relatively low quantitative cancer risk estimates. Given these low estimates and the uncertainty associated with classifying perc, it is not surprising that the EPA had difficulty justifying the establishment of a national emission standard. More recently, quantitative risk analysis has played an important role in keeping the industry involved in the voluntary DfE program. However, it was not a risk assessment or cost-benefit analysis itself that motivated the industry, but the threatened release of a quantitative risk assessment.

Under the Clean Air Act Amendments of 1990, the national emission standard was based on technology instead of risk. However, several years after the establishment of the technology standard, the EPA must perform "residual risk" assessments to determine whether further controls are warranted. Thus the EPA's risk assessment activities for perc dry cleaners are likely to continue.

Given the equivocal scientific findings on perc's human carcinogenicity, the use of risk assessment as a basis for regulation has led to substantial expenditures for development of regulations and for scientific research. The issue of whether or not these resources have been well spent is certainly debatable, although it is impossible to know with certainty how the resources would have otherwise been spent and it seems unlikely that we will ever definitively know whether exposure to perc causes significant adverse health effects in humans.

| Sources

Anderson, E., 1985. Memo from Elizabeth Anderson, Director, Office of Health and Environmental Assessment, to Joseph Cotruvo, Director, Criteria and Standards Division; John O'Connor, Director, Strategies and Air Division; and Don Clay, Director, Office of Toxic Substances, EPA, September 29.

Aschengrau, A., D. Ozonoff, C. Paulu, et al., 1993. "Cancer Risk and Tetrachloroethylene-Contaminated Drinking Water in Massachusetts," *Archives of Environmental Health* 48(5):284-292.

Agency for Toxic Substances and Disease Registry (ATSDR), 1993. *Toxicological Profile for Tetrachloroethylene.* ATSDR/TP-92/18 (Atlanta, Georgia: Department of Health and Human Services, Agency for Toxic Substances and Disease Registry).

Blair, A., P. Decoufle, and D. Grauman, 1979. "Causes of Death Among Laundry and Dry Cleaning Workers," *American Journal of Public Health* 69:508-511.

Brown, D.P. and S.D. Kaplan, 1985. *Retrospective Cohort Mortality Study of Dry Cleaning Workers Using Perchloroethylene.* (Cincinnati, Ohio: Department of Health and Human Services, National Institute for Occupational Safety and Health).

Cavalleri, A., F. Gobba, M. Paltrinieri, et al., 1994. "Perchloroethylene Exposure Can Induce Colour Vision Loss." *Neuroscience Letters* 179:162-166.

California Air Resources Board (CARB), 1991. *Proposed Identification of Perchloroethylene as a Toxic Air Contaminant, Technical Support Documents* (August 1991) and revisions (April 1992), (Sacramento, Calif.: California Air Resources Board and the Department of Health Services).

Center for Emissions Control (CEC), 1992. *Dry Cleaning: An Assessment of Emissions Control Options.* (Washington, D.C.: Center for Emissions Control).

Chemical Marketing Reporter (CMR), 1976. "Chemical Profile: Perchloroethylene," August 9.

———, 1978. "Perc Injunction Won by Industry; CPSC is Loser" September 18.

———, 1979. "Chemical Profile: Perchloroethylene," June 18.

Chemical Regulation Reporter (CRR), 1980. "Scientific Review Board Counters EPA Estimates of Chemical Cancer Risks," September 12.

Department of Labor (DOL), 1938. *Job Descriptions for the Cleaning, Dyeing, and Pressing Industry.* (Washington, D.C.: U.S. Department of Labor and U.S. Employment Service).

———, 1993. "Permissible Exposure Levels for 376 Toxic Substances Vacated." *Labor News,* March 23, U.S. Department of Labor, Occupational Safety and Health Administration, Office of Information, press release USDL 93-103.

Doull, J., 1988. Letter from John Doull, Chair, Halogenated Organics Subcommittee; Richard Greisemer, Chair, Environmental Health Committee; and Norton Nelson, Chair, SAB Executive Committee, to Lee Thomas, Administrator, EPA, March 9.

Drye, E., 1991. "Perchloroethylene" in J.D. Graham, *Harnessing Science for Environmental Regulation.* (New York: Praeger), pp. 97-125.

Duh, R.W. and N.R. Asal, 1984. "Mortality Among Laundry and Dry-Cleaning Workers in Oklahoma," *American Journal of Public Health* 74(11):1278-1280.

Echeverria, D., R.F. White, and C. Sampaio, 1995. "A Behavioral Evaluation of PCE Exposure in Patients and Dry Cleaners: A Possible Relationship Between Clinical and Preclinical Effects," *Journal of Occupational and Environmental Medicine* 37(6):667-680.

Environmental Protection Agency (EPA), 1977. "Air Quality: Recommended Policy on Control of Volatile Organic Compounds." *Federal Register* 42:35314-35316, July 8 (Washington, D.C.: Environmental Protection Agency).

———, 1978. *Control of Volatile Organic Emissions from Perchloroethylene Dry Cleaning Systems,* EPA-450/2-78-050 (Washington, D.C.: Environmental Protection Agency).

———, 1980. "Standards of Performance for New Stationary Sources; Perchloroethylene Dry Cleaners. Proposed Rule." *Federal Register* 45:78174-78181, November 25 (Washington, D.C.: Environmental Protection Agency).

———, 1983. "Air Quality: Proposed Revision to Agency Policy Concerning Ozone SIP's and Solvent Reactivities." *Federal Register* 48:49097-49098, October 24 (Washington, D.C.: Environmental Protection Agency).

———, 1984a. "Draft Health Assessment Document for Tetrachloroethylene (Perchloroethylene)." *Federal Register* 49:10575-10576, March 21 (Washington, D.C.: Environmental Protection Agency).

———, 1984b. "Draft Criteria Document for Tetrachloroethylene," February (Washington, D.C.: Environmental Protection Agency, Office of Drinking Water).

———, 1985a. *Health Assessment Document for Tetrachloroethylene (Perchloroethylene), Final Report,* EPA/600/8-82/005F (Washington, D.C.: Environmental Protection Agency).

———, 1985b. "Assessment of Perchloroethylene as a Potentially Toxic Air Pollutant. Proposed Rule." *Federal Register* 50:52880-52884, December 26 (Washington, D.C.: Environmental Protection Agency).

———, 1986a. *Addendum to the Health Assessment Document for Tetrachloroethylene (Perchloroethylene), Updated Carcinogenicity Assessment,* EPA/600/8-82/005FA (Washington, D.C.: Environmental Protection Agency).

———, 1986b. "Guidelines for Carcinogen Risk Assessment." *Federal Register* 51:33992-34003, September 24 (Washington, D.C.: Environmental Protection Agency).

———, 1986c. Memorandum from Robert L. Ajax, Chief of Standards Development Branch, and Susan R. Wyatt, Chief of Chemicals and Petroleum Branch, to Jack R. Farmer, Director of Emission Standards and Engineering Division, EPA, August 28.

———, 1986d. "Hazardous Waste Management System: Land Disposal Restrictions. Proposed Rule." *Federal Register* 51:1602, January 14 (Washington, D.C.: Environmental Protection Agency).

———, 1986e. "Hazardous Waste Management System: General; Identification and Listing of Hazardous Waste; Standards for Generators of Hazardous Waste; Standards for Transporters of Hazardous Waste; EPA Administered Permit Programs; Authorization of State Hazardous Waste Programs. Final rule." *Federal Register* 51:10146-10176, March 24 (Washington, D.C.: Environmental Protection Agency).

———, 1987. "EPA Staff Comments on Issues Regarding the Carcinogenicity of Perchloroethylene (Perc) Raised by the SAB," July 30 (Washington, D.C.: Environmental Protection Agency).

———, 1988. *Options for Regulating Perchloroethylene Emissions in the Dry Cleaning Industry: A Cost-Benefit Analysis, Draft Report.* January (Washington, D.C.: Environmental Protection Agency), (unpublished draft on file with the author).

———, 1989. "Hazardous Waste Management System: Identification and Listing of Hazardous Waste CERCLA Hazardous Substance Designation; Reportable Quantity Adjustment. Final rule." *Federal Register* 54:50968-50977, December 11 (Washington, D.C.: Environmental Protection Agency).

———, 1991a. *Dry Cleaning Facilities: Background Information for Proposed Standards, Draft EIS,* EPA-450/3-91-020a (Washington, D.C.: Environmental Protection Agency).

———, 1991b. *Economic Impact Analysis of Regulatory Controls in the Dry Cleaning Industry, Final Report,* EPA-450/3-91-021 (Washington, D.C.: Environmental Protection Agency, Office of Air Quality Planning and Standards).

———, 1991c. *Response to Issues and Data Submissions on the Carcinogenicity of Tetrachloroethylene (Perchloroethylene),* EPA/600/6-91/002F (Washington, D.C.: Environmental Protection Agency, Office of Research and Development).

———, 1991d. "Standards of Performance for New Stationary Sources; Perchloroethylene Dry Ceaners. Proposed Rule. Withdrawal." and "National Emission Standards for Hazardous Air Pollutants for Source Categories: Perchloroethylene Dry Cleaning Facilities. Proposed Rule." *Federal Register* 56:64382-64402, December 9 (Washington, D.C.: Environmental Protection Agency).

———, 1992a. "Air Quality: Revision to Definition of Volatile Organic Compounds. Proposed Rule." *Federal Register* 57:48490-48492, December 9 (Washington, D.C.: Environmental Protection Agency).

———, 1992b. *Proceedings: International Roundtable on Pollution Prevention and Control in the Drycleaning Industry, May 27-28,* EPA/774/R-92/002 (Washington, D.C.: Environmental Protection Agency).

———, 1992c. "Report on Perchloroethylene: T.S.C.A. Docket, R.M. 1 Dossier," Charlie Auer, Chairman, EPA Office of Pollution Prevention and Toxics, June 3.

———, 1993. "National Emission Standards for Hazardous Air Pollutants for Source Categories: Perchloroethylene Dry Cleaning Facilities. Final rule." *Federal Register* 58:49354-49380, September 22 (Washington, D.C.: Environmental Protection Agency).

———, 1994a. *Revisions to the Guidelines for Carcinogen Risk Assessment, External Review Draft,* EPA/600/BP-92/003 (Washington, D.C.: Environmental Protection Agency).

———, 1994b. "Protection of Stratospheric Ozone. Final rule." *Federal Register* 59:13044-13146, March 18 (Washington, D.C.: Environmental Protection Agency).

———, 1995. "The DfE Dry Cleaning Project," in *Design for the Environment:*

Building Partnerships for Environmental Improvement, EPA/600/K-93/002 (Washington, D.C.: Environmental Protection Agency, Office of Pollution Prevention and Toxics).

———, 1996a. *Proposed Guidelines for Carcinogen Risk Assessment.* (Washington, D.C.: Environmental Protection Agency). Office of Research and Development, EPA/600/P-92/003C.

———, 1996b. "Air Quality: Revision of Definition of Volatile Organic Compounds — Exclusion of Perchloroethylene. Final Rule." *Federal Register* 61:4588-4591, February 7 (Washington, D.C.: Environmental Protection Agency).

Hattis, D., S. Tuler, L. Finkelstein, and Z. Luo, 1986. *A Pharmacokinetic/Mechanism-Based Analysis of the Carcinogenic Risk for Perchloroethylene.* (Cambridge, Massachusetts: MIT Center for Technology, Policy and Industrial Development).

International Fabricare Institute (IFI), 1990. "Maintenance of Water Separators and Disposal of Separator Water," July (Silver Spring, Maryland: International Fabricare Institute).

———, 1995. "The Small Business Fabric Care Superfund Coalition Reauthorization Proposal," July 7.

International Trade Commission (ITC), 1960-1993. *Synthetic Organic Chemicals, United States Production and Sales,* published annually (Washington, D.C.: International Trade Commission)

Jehassi, O., 1995. Interview with Ohad Jehassi, EPA Office of Pollution Prevention and Toxics, December 5.

Johnson, A.E., 1971. *Drycleaning.* Monograph MM/TT/6 (Watford, England: Merrow Publishing Co. Ltd.), pp 1-7.

Kaplan, S.D., 1980. *Dry Cleaning Workers Exposed to Perchloroethylene. A Retrospective Cohort Mortality Study,* Contract No. 210-77-0094 (Cincinnati, Ohio: Department of Health and Human Services, National Institute for Occupational Safety and Health).

Katz, R.M. and D. Jowett, 1981. "Female Laundry and Dry-Cleaning Workers in Wisconsin. A Mortality Analysis." *American Journal of Public Health* 71:305-307.

Lake, M.R., 1992. Letter from Michael R. Lake, Air Pollution Control District, County of San Diego, to George Smith, EPA, January 8.

Lin, R.S. and I.I. Kessler, 1981. "A Multifactorial Model for Pancreatic Cancer in Man." *Journal of the American Medical Association* 245:147-152.

Loehr, R.L., 1991. Letter from Raymond Loehr, Chair, Science Advisory Board, and Bernard Weiss, Acting Chair, Environmental Health Committee, to William Reilly, EPA Administrator, August 16.

Lowrance, S.K., 1993. Letter from Sylvia K. Lowrance to William Fisher, International Fabricare Institute, June 2.

Lyle, D.S., 1977. *Performance of Textiles.* (New York: John Wiley and Sons, Inc.), p 326.

Meijer, J., 1994. Interview with Jon Meijer, International Fabricare Institute, August 18.

Mutti, A., R. Alinovi, E. Bergamaschi, et al., 1992. "Nephropathies and Exposure to Perchloroethylene in Dry-Cleaners." *Lancet* 25(340):189-193.

National Cancer Institute (NCI), 1977. *Bioassay of Tetrachloroethylene for Possible Carcinogenicity*, DHEW Pub. No. (NIH) 77-813 (Bethesda, Maryland: Department of Health, Education, and Welfare, National Institutes of Health).

National Toxicology Program (NTP), 1986. *Toxicology and Carcinogenesis of Tetrachloroethylene (Perchloroethylene) (CAS No. 127-18-4) in F344/N Rats and B6C3F1 Mice (Inhalation Studies)*, NIH Publication No. 86-2567, Technical Report Series 311 (Bethesda, Maryland: National Institutes of Health, National Toxicology Program).

New York State Department of Health (NYDOH), 1991. *Investigation of Indoor Air Contamination in Residences Above Dry Cleaners*, October (New York: Department of Health, Bureau of Toxic Substances, and Department of Environmental Conservation, Division of Air Resources). Reprinted under same title in Schreiber J.S., et al. (1993) *Risk Analysis* 13(3):335-344.

Occupational Health and Safety Administration (OSHA), 1988. "Air Contaminants. Proposed Rule." *Federal Register* 53:20960, June 7 (Washington, D.C.: Department of Labor, Occupational Health and Safety Administration).

———, 1989. "Air Contaminants. Final Rule." *Federal Register* 54:2686-2689, January 19 (Washington, D.C.: Department of Labor, Occupational Health and Safety Administration).

———, 1992. "Air Contaminants. Proposed Rule." *Federal Register* 57:26002-26416, June 12 (Washington, D.C.: Department of Labor, Occupational Health and Safety Administration).

Portman, R., M. Frost, J. Christensen, and J. Ramstad, 1995. Letter to Carole Browner, EPA Administrator, February 28.

Price, R.G., S.A. Taylor, E. Crutcher, et al., 1995. "The Assay of Laminin Fragments in Serum and Urine as an Indicator of Renal Damage Induced by Toxins." *Toxicology Letters* 77:313-318.

Professional Wet Cleaning Partnership (PWCP), 1995. Letter of Agreement signed by representatives of the International Fabricare Institute, Neighborhood Cleaners Association, Fabricare Legislative and Regulatory Education Council, Federation of Korean Drycleaning Associations, Greenpeace, Center for Neighborhood Technology, Massachusetts Toxics Use Reduction Institute, and Union of Needletrades, Industrial and Textile Employees, November 1.

Rampy, L.W., J.F. Quast, M.F. Balmer, et al., 1978. *Results of a Long-Term Inhalation Toxicity Study on Rats of a Perchloroethylene (Tetrachloroethylene) Formation*, (Midland, Michigan: Dow Chemical, Co., Toxicology Research Laboratory).

Rhomberg, L., 1986. "Human Cancer Risks from Perchloroethylene." Memo from Lorenz Rhomberg to J. William Hirzy through Joseph A. Carra, Office of Pesticides and Toxic Substances, EPA, January 8.

Rice, B., 1994. Letter from Bonnie Rice, Greenpeace, to Mark Greenwood, Director of EPA OPPT, August 10.

Risotto, S., 1994. Interview with Steve Risotto, Center for Emissions Control, August 17.

———, 1995. Interview with Steve Risotto, Center for Emissions Control, December 8.

Ruder, A.M., E.M. Ward, and D.P. Brown, 1994. "Cancer Mortality in Female and Male Dry-Cleaning Workers." *Journal of Occupational Medicine* 36(8):867-874.

Seitz, B., 1994. Interviews with Bill Seitz, Neighborhood Cleaner's Association, August 18 and September 21.

Sheiman, D., 1992. Letter from Deborah A. Sheiman, Natural Resources Defense Council, to William Johnson, EPA, December 28.

Spaw, S., 1990. Letter from Steve Spaw, Executive Director of the Texas Air Control Board, to Bob Craig, International Fabricare Institute, October 26.

Spedden, S.E. and P.B. Ryan, 1992. "Probabilistic Connotations of Carcinogen Hazard Classifications: Analysis of Survey Data for Anchoring Effects." *Risk Analysis* 12(4):535-541.

SRI International (SRI), 1994. "C2 Chlorinated Solvents," *Chemical Economics Handbook,* (Menlo Park, Calif.: SRI International).

Street , A.L.H., 1926. *Law for Cleaners and Dyers and Laundryowners.* (New York: Dowst Publishing Co.), pp 13-47.

Theiss, J.C., G.D. Stoner, M.D. Shimkin, and E.K. Weisburger, 1977. "Tests for Carcinogenicity of Organic Contaminants of United States Drinking Waters by Pulmonary Tumor Response in Strain A Mice," *Cancer Research* 37:2717-2720.

Thomas, L., 1987. Letter from Lee Thomas, EPA Administrator, to Norton Nelson, Chair, SAB Executive Committee, August 3.

Thomas, K.W., E.D. Pellizzari, R.L. Perritt, and W.C. Nelson, 1991. "Effect of Dry-Cleaned Clothes on Tetrachloroethylene Levels in Indoor Air, Personal Air, and Breath for Residents of Several New Jersey Homes." *Journal of Exposure Analysis and Environmental Epidemiology* 1(4):475-490.

Tichenor, B.A., L.E. Sparks, M.D. Jackson, et al., 1990. "Emissions of Perchloroethylene from Dry Cleaned Fabrics." *Atmospheric Environment* 24A(5):1219-1229.

Van Duuren, B.L., M. Goldschmidt, G. Loewengart, et al., 1979. "Carcinogenicity of Halogenated Olefinic and Aliphatic Hydrocarbons in Mice." *Journal of the National Cancer Institute* 63:1433-1439.

Voorhees, S., 1989. "Development of Proposed H.R. 3030 Title III Pollutant List." Memo from Scott Voorhees, EPA Program Analysis and Technology Section, to Stanley A. Meiburg, EPA Planning and Management Staff, September 5.

Voytek, P., 1994. Letter from Peter Voytek, Halogenated Solvents Industry Alliance, to Jean Elizabeth Parker, EPA, Design for the Environment Program Staff Director, June 27.

Wang, C.J., D.R. Blake, and F.S. Rowland, 1995. "Seasonal Variations in the Atmospheric Distribution of a Reactive Chlorine Compound, Tetrachloroethylene (CCl_2=CCl_2)." *Geophysical Research Letters* 22(9): 1097-1100.

Weiss, N.S., 1995a. "Cancer in Relation to Occupational Exposure to Perchloroethylene." *Cancer Cases and Control* 6:257-266.

———, 1995b. "Observers Report on the Meeting of the IARC Working Group on the Evaluation of Carcinogenic Risks to Humans, Volume 63: Dry Cleaning, Some Solvents and Other Industrial Chemicals," unpublished notes prepared by Noel S. Weiss, University of Washington, March 21.

Wentz, M., 1995. "What's New About Petroleum Cleaning Technology?" *American Drycleaner* (November):85.

Wilkin, J.G., K.C. Cheng, P. Tam, et al., 1995. *The Canadian Dry Cleaning Sector. Part II: Assessment of Alternative Technologies,* Report prepared for Environment Canada, Commercial Chemicals Division, October (Richmond, British Columbia: B.H. Levelton & Associates, Ltd.)

Wolf, K. and C.W. Myers, 1987. *Hazardous Waste Management by Small Quantity Generators: Chlorinated Solvents in the Dry Cleaning Industry,* Report R-3505-JMO/RC (Santa Monica, Calif.: Rand Corporation).

Wolf, K., 1992. "Case Study: Pollution Prevention in the Dry Cleaning Industry: A Small Business Challenge for the 1990s," *Pollution Prevention Review* (Summer):311-330.

FIVE

Fewer Fumes from Coke Plants

Jennifer Kassalow Hartwell
John D. Graham

Steel plays a vital role in the construction of products ranging from automobiles and appliances to agricultural machinery and military hardware. At the same time, steel production is a persistent source of environmental problems, including conventional air pollution, surface water pollution, disposal of hazardous wastes, and emissions of carbon dioxide and other "greenhouse gases."

In steel-producing towns such as Pittsburgh, Pennsylvania and Birmingham, Alabama, many mills have been closed since the industry's peak production levels in the 1970s. But among the mills that have survived foreign competition, recessions, and cost pressures, substantial progress has been made in curbing pollution. Many steel-producing regions of the country are now achieving compliance with the U.S. Environmental Protection Agency's health-based air quality standards for particulate matter and sulfur dioxide. However, there are concerns that steel plants remain a significant source of toxic air pollution.

In this chapter, we examine recent efforts by the EPA and the steel industry, under the Clean Air Act Amendments of 1990, to solve a persistent problem: so-called fugitive emissions of hazardous gases and particles from coke production, the operation in integrated steelmaking that has proven to be the most difficult to control. For residential areas located in close proximity to steel plants, it is typically the coke plant that is the source of community complaints and advocacy efforts by grassroots groups.

From a risk management perspective, the recent regulation of coke plants under Title III of the Clean Air Act Amendments (CAAA) of 1990 is paradoxical. Title III pertains to so-called hazardous air pollutants and "coke oven emissions" are one of the 189 listed hazardous air pollutants. The EPA's quantitative estimates of cancer risk associated with coke oven emissions played an important role in making the case for tighter controls on coke plants. Yet Congress ultimately decided to regulate this industry through technology-based standards, followed only by the possibility of risk-based standards. Coke plant operators may elect a more stringent set of technology-based standards and thereby avoid a risk-based standard until 2020. Congress proceeded with this approach even though the EPA's most recent economic analysis suggests that costs will be large compared with health benefits. This chapter traces the role of risk assessment in the EPA's regulatory program for coke plants and concludes with a preliminary assessment of how owners of coke plants are responding to the command-and-control rules developed by the EPA through "regulatory negotiation."

| The Coke-Steel Connection

Each year the United States consumes more than 100 million tons of steel in the manufacture of automobiles, agricultural machinery, appliances, and other products. Although coke-free methods of steel production are emerging, coke-based steelmaking remains dominant throughout the world and is projected to maintain its dominance into the twenty-first century.

Coke-based steelmaking involves three basic steps: the preparation of raw materials, including coke, limestone, and iron ore; use of those materials in the blast furnace to produce pig iron; and creation of molten raw steel from the pig iron and scrap in the basic oxygen furnace. Coke, which is the residue from the destructive distillation of coal, plays a critical part in the production of steel because it has important physical properties. Unlike coal, coke has great strength, which enables it to withstand breakage during handling and use in the blast furnace. Since it is nearly all carbon, coke also burns more efficiently and at far higher temperatures than coal.

In the United States, coke is produced by two types of businesses. Merchant coke producers are independently owned companies that produce coke for sale on the open market. These producers rely on coke as their primary source of revenue and in 1990 were responsible for approximately 15 percent of the coke produced in the United States. The other 85 percent of U.S. coke is produced by integrated steel producers. Unlike merchant producers, integrated steel firms are primarily concerned with producing molten iron, molten steel, and finished steel. For these firms, coke is produced as a means to an end (though some integrated firms also sell coke on the open market).

| Making Coke

The first step in coke-based steelmaking is the production of coke. "By-product cokemaking," in which the volatile material derived from coal during the coking process is collected, was first used in the American steel industry in 1895 and remains today the most economical method of making coke. The process, illustrated in Figure 5.1, begins as coal is received at plants via rail, river barge, or, occasionally, truck. The coal is then blended, crushed, and screened before it is transferred to a coal storage bunker. The prepared coal is loaded from the storage bunker into a "larry" car that travels the length of the "coke battery" — as many as 100 ovens, each separated by long and narrow brick-walled flue chambers (Lankford et al., 1985). The oven doors are closed and sealed before an oven is "charged" with coal from the larry car through the coal port at the top of the oven.

The coal is then heated at over 1000 degrees Centigrade for approximately eighteen hours, during which time the volatile and nonvolatile components of the coal are separated. The coke oven gas containing volatilized materials (such as tar, benzene, and toluene) is directed to a by-product plant where some chemicals are recovered for commercial use and others are used to reheat the battery and fuel other parts of the steelmaking process. Occasionally the coke oven gas is sold to electric utilities. Once the coking process is completed, the hot coke is pushed out of the oven slots, transported by larry car to a quenching tower for cooling with water, and transported to a blast furnace, where pig iron is made.

| Coke in the Blast Furnace

The traditional method of making steel entails a blast furnace, a large steel vessel lined with insulating brick. Iron ore mixed with limestone is added from the top in alternating layers with coke. Heated air, which is

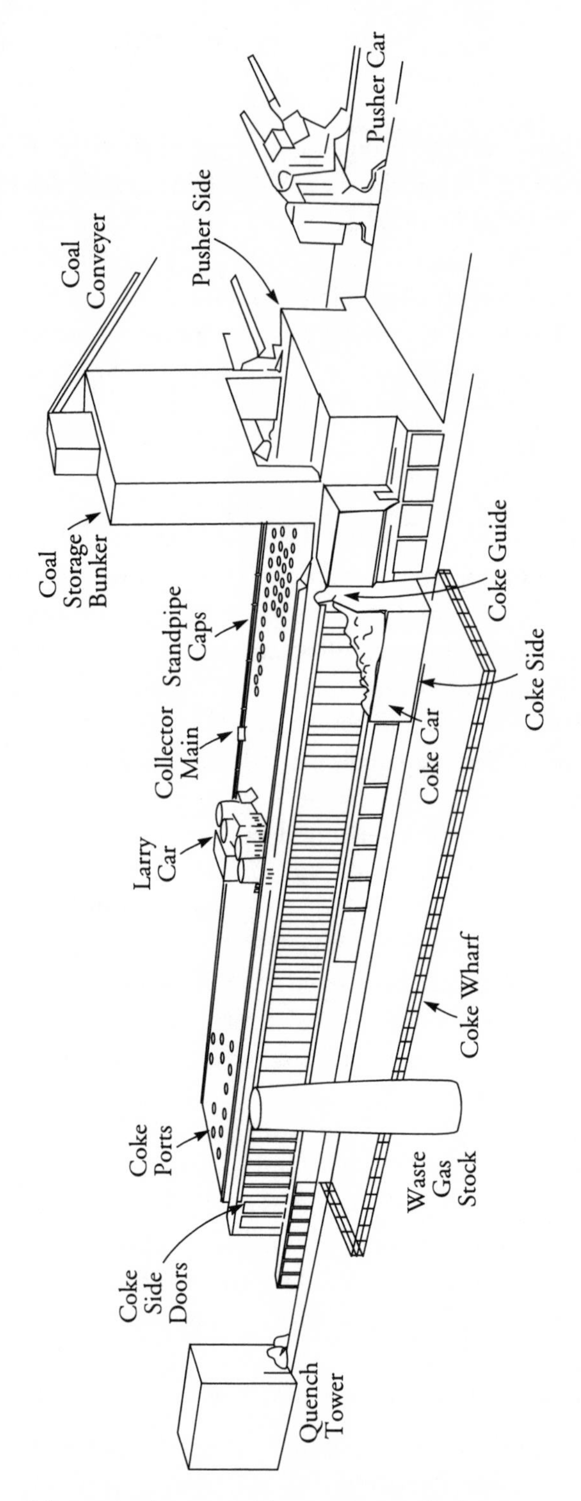

FIGURE 5.1 | Schematic diagram of by-product coke battery.

blown in from the bottom to fuel the burning coke, melts the iron ore and supplies carbon to transform ore from iron oxide to liquid iron. The liquid iron is taken to another furnace, a basic oxygen furnace, where it is combined with scrap and injected with oxygen to reduce the carbon content, thereby producing steel (Holusha, 1994b).

The main function of coke is to provide the carbon used to reduce iron ore to iron. In addition coke produces heat, and because the coke retains its strength at temperatures above the melting temperature of pig iron and slag, it provides the structural support that keeps the unmelted burden materials from falling into the hearth (Lankford et al., 1985:543).

Air Pollution at Coke Plants

Although most byproducts of coke production are recovered, air pollution can occur at several points. During the coking process itself, fugitive emissions can escape from door linings, lids, and cracks. Doors are currently the single largest source of fugitive emissions from the coke battery. Gas and particles can escape into the air when the coke battery doors are opened and closed, and from leaks in closed doors.

If the coal is heated for a sufficient length of time, the pushing operation poses little problem. However, if the coal (not yet coke) is prematurely pushed from the oven, flames and large quantities of volatile gases shoot out from the oven. This is particularly likely to occur in older ovens, which may, due to damaged flues, fail to heat the coal uniformly. Consequently only a portion of the charge is completely converted to coke. This sort of incident, called a "green push," often occurs when coke plant operations are not under proper control; they are a nightmare for operators, because batteries can be damaged, workers may be hurt, and there will more likely be complaints from neighboring communities.

Basis of Human Health Concerns

Coke ovens emit a complex mixture of gases and particles that can be toxic and carcinogenic (EPA, 1984). The mixture includes literally hundreds of chemicals, including polycyclic organic matter from coal tar pitch volatiles, beta-naphthylamine, benzene, beryllium, and compounds of arsenic, cadmium, and lead. Not all of the constituents making up the mixture have been tested for toxicity or even identified.

Since the 1960s, workers, labor unions, environmental groups, and local health departments have sought stringent control of fugitive emissions from by-product coke ovens. The area around Pittsburgh suffered from pollution from USX Corporation's Clairton Works, the largest coking facility in the country, as well as several smaller coke plants run by other steelmakers. At one time, community groups claimed that the stench, haze, and dirt produced by the pollutants affected the area's ability to attract top professionals and resulted in a drop in property values and a general decline in the quality of life in several communities around Pittsburgh (Goldburg, 1989). Consequently, Pennsylvania's Allegheny County Health Department emerged as one of the earliest advocates of emission control policies.

The highly toxic benzene soluble organics (BSOs) component of coke oven emissions have been singled out for intensive study and standard setting. Coal tar aerosols in air samples taken near coke ovens have been shown to cause lung cancer in laboratory animals and to test positive in mutagenicity tests conducted on bacteria. Urine samples from nonsmoking coke oven workers have also been deemed mutagenic. Risks of nonfatal cancer and other potential health effects may also exist, but no numerical expressions of these risks have been developed (EPA, 1987: E-6).

Epidemiological studies have provided evidence of the carcinogenic effects of large exposures to coke oven emissions. Several studies have found that coke oven workers suffer from a significantly raised risk of fatal respiratory cancer. In fact, employees who worked on top of the ovens were found to have a risk of lung, trachea, and bronchus cancer seven times higher than nonoven coke production workers (Lloyd, 1971).

| Pre-1990 Progress in Controlling Air Emissions

Control of air pollution from coke plants did not begin with the Clean Air Act Amendments of 1990. In fact, emissions control began inadvertently in the early 1900s during the transition from the beehive technology, which allowed most of the gases generated during the coking process to escape into the atmosphere, to the contemporary by-product technology. Beehive technology was replaced with by-product ovens due to the rising prices of certain chemical products in World War I and the commercial desire to capture and sell the byproducts of coke production. The coke industry began to prosper after 1910 when war threatened supplies of chemicals from Germany. For instance, benzol was needed to upgrade the octane value in gasoline and phenol for plastics. The American steel industry soon found that selling by-products from coke production was a profitable venture.

Once steel companies, mostly dependent on beehive technology until 1920, realized the profit-making potential of coke by-products, as well as the use of coke oven gas for fueling their coke ovens and blast furnaces, they began to build their own coke production facilities (Anderson, 1990). Normally 25 to 40 percent of coal by weight is converted into by-products in the by-product coke oven, with the majority of the coal (60 to 75 percent) being converted into coke. However, by-products became so valuable that it was later worth converting 50 percent of the coal by weight into by-products instead of the normal 25 to 40 percent (Farley, 1994).

In recent decades the role of the by-product coke plant has changed. The by-products from coke production have met intensive competition from a growing petrochemical industry. Instead of being viewed as a big money maker, by-product plants began to be valued for their "smooth" production — their effectiveness at removing tar and other "gunk" from the system.

Prior to the passage of the Occupational Safety and Health Act of 1970, a performance standard (threshold limit value) for coke emissions was promulgated under the Walsh-Healey Public Contracts Act. In the 1960s and 1970s, human exposures to emissions from the "charging" and "pushing" operations were reduced through improvements in work practices, technological advances, and wider use of respirators in the workplace. Later, concerns about particle pollution from by-product coke ovens, like those expressed by citizens in Allegheny County, Pennsylvania, stimulated environmental regulation of coke plants. And since many coke plants were located in towns not meeting the EPA's health-based standard for particulates in outdoor air, in the 1970s the EPA and states began to compel emissions control at coke plants as part of their efforts to meet the EPA's particulate standard.

| The 1970 Clean Air Act: Zero Risk?

Under Section 112 of the Clean Air Act of 1970, the EPA is required to protect public health from toxic emissions. Section 112 (a)(1) of the 1970 law defined a "hazardous" air pollutant as "an air pollutant to which no ambient air quality standard is applicable and which in the judgment of the Administrator causes, or contributes to, air pollution which may reasonably be anticipated to result in an increase in serious irreversible, or incapacitating reversible, illness." Once a pollutant is deemed hazardous by the EPA administrator, the administrator must set an allowable emissions standard based on the level that she or he determines protects the public health with an "ample margin of safety."

The EPA got off to a slow start with this new responsibility. The agency felt it had been ordered to follow "hopelessly unrealistic" deadlines (Percival et al., 1992), estimating that it would take three to seven years to complete each risk assessment, the reports necessary for supporting a final emissions standard for a substance. Congress had given the EPA a total of only fifteen months not only to identify hazardous pollutants but also to propose and issue final regulations. More important, the agency feared the policy implications of the generally accepted no-threshold assumption of carcinogens — an assumption that led some experts to conclude that the only level of carcinogenic emissions that would provide "an ample margin of safety" was zero.

A zero-emissions standard would most definitely require the shutdown of major industries, including the steel and chemical manufacturing industries. The EPA's reluctance to confront this interpretation contributed to its slow progress in proposing standards for pollutants. By 1976 the EPA had proposed standards for only four pollutants (beryllium, mercury, asbestos, and vinyl chloride), and these were issued only after a court action was initiated by the Environmental Defense Fund (Percival et al., 1992). It took the EPA almost a decade to decide whether to formally list coke oven emissions under Section 112, especially since the agency's Science Advisory Board was skeptical of the EPA's quantitative approach to the estimated cancer risk of coke oven emissions.

| The Risks of Coke Plants

In the 1980s, fugitive emissions from oven doors during the coking process became a primary concern of community groups, the Natural Resources Defense Council (NRDC), and regulatory officials. Consequently, in 1987 the EPA listed coke oven emissions as a hazardous air pollutant and released its first quantitative risk assessment of coke plants.

There was no direct evidence of cancer risk to residents living near coke facilities, although one later study suggested that pollution from steel mills as a whole is associated with health problems in surrounding communities (Pope, 1991). Lacking direct data on the community health impacts of coke oven emissions, in 1987 the EPA risk assessors projected cancer risks in the communities near coke plants based on the excess cancer risks experienced by coke oven workers (Graham, 1993: 12). The EPA's risk assessment suggested that the cancer risk caused by the entire industry (up to 3 additional cases of fatal cancer per year nationwide) was small relative to the 500,000 fatal cancers recorded each year. But even so, this modest cancer risk was larger than those typically estimated by the EPA for many other industries. Moreover, the "maximally exposed" residents near coke plants were estimated to suffer substantial elevations in lifetime risk of respiratory cancer — as high as a 1 in 50 to 1 in 1,000 chance of developing cancer after a lifetime of exposure (EPA, 1987) (Table 5.1). Since many EPA programs were using 1 chance in 1,000,000 as an indication of negligible risk, agency officials became convinced that coke plants were in need of further regulation (Rosenthal, Gray, and Graham, 1992).

Questions were raised about the accuracy of the EPA's risk estimates. One key assumption was that the dose-response relationship for cancer formation is linear as BSO exposures decline, even at exposure levels well below those experienced by coke oven workers (Rueter and Steger, 1990: 51). Several nonlinear models, which also fit the occupational data well, produce much smaller estimates of risk for community residents (Lamm, 1987). The EPA's Science Advisory Board urged the agency to present several alternative risk models, but the EPA placed primary emphasis on the results of its standard linear model.

Human exposure estimates were also uncertain. Direct measurements of coke oven emissions are usually infeasible, due to the fugitive

TABLE 5.1 | Coke oven maximum individual risk estimates, 1989

PLANT	LOCATION	MAXIMUM INDIVIDUAL RISK
ABC Coke	Tarrant, AL	9.9 in 1,000
Empire Coke	Holt, AL	4.5 in 1,000
Koppers	Woodward, AL	1.7 in 100
Gulf States Steel	Gadsden, AL	9.9 in 1,000
Sloss Industries	Birmingham, AL	9.1 in 1,000
USX	Fairfield, AL	2.6 in 10,000
National Steel	Granite City, IL	1,6 in 1,000
Acme Steel	Chicago, IL	2.0 in 1,000
LTV Steel	So. Chicago, IL	6.0 in 10,000
Bethlehem Steel	Burns Harbor, IN	5.6 in 10,000
Citizens Gas	Indianapolis, IN	5.1 in 1,000
Welsh Coke	Terre Haute, IN	3.3 in 1,000
Inland Steel	E. Chicago, IN	1.4 in 100
USX	Gary, IN	1.1 in 100
Armco Inc.	Ashland, KY	2.6 in 1,000
Bethlehem Steel	Ecorse, MI	2.0 in 10,000
Rouge Steel	Dearborn, MI	9.3 in 10,000
National Steel	Ecorse, MI	1.7 in 10,000
Tonawanda	Buffalo, NY	8.6 in 10,000
Bethlehem Steel	Lackawanna, NY	1.7 in 10,000
LTV Steel	Warren, OH	1.6 in 1,000
Armco Inc.	Middletown, OH	7.4 in 1,000
New Boston	Portsmouth, OH	2.2 in 1,000
Toledo Coke	Toledo, OH	1.5 in 1,000
LTV Steel	Cleveland, OH	4.3 in 1,000
USX	Lorain, OH	8.2 in 1,000
Bethlehem Steel	Bethlehem, PA	1.1 in 100
LTV Steel	Pittsburgh, PA	8.3 in 1,000
Erie Coke	Erie, PA	2.1 in 10,000
Shenango	Pittsburgh, PA	1.3 in 1,000
USX	Clairton, PA	1.8 in 100
Sharon Steel	Monessen, PA	1.6 in 1,000
Lone Star Steel	Lone Star, TX	4.1 in 10,000
Geneva Steel	Provo, UT	1.1 in 100
Wheeling-Pitt	E. Steubenville, WV	6.9 in 1,000
Detroit Coke	Detroit, MI	1.0 in 1,000

Source: Environmental Protection Agency, Office of Air Quality Planning, Research Triangle Park, North Carolina.

nature of the emissions. It is difficult to assess the number of emission points per battery and the emission rate per emission point (CAAA, 1990: 340–341). Emission rates used by the EPA are often educated guesses based on inspection of a limited number of batteries and disjointed measurements (Graham, 1993: 17). The validity of the predicted air concentration in the community, produced by a dispersion model, was also of concern. The EPA found that the maximum, long-term BSO concentrations predicted by its more rudimentary dispersion model were three to five times larger than the predictions from a more refined site-specific model applied to two coke plants (EPA, 1987: E-15).

When the EPA calculated the risk to the so-called maximally exposed individual, it assumed that a resident lived at the point of maximum modeled air concentration of BSOs, usually 200 meters from the plant. Sometimes residences are located immediately downwind of and next to a plant fence line, but this is not always true. In other cases the maximum concentration point is not in a residential neighborhood but in farmland, industrial land, or above bodies of surface water. Modeled pollution concentrations decline rapidly with distance from the source. Therefore, the estimated exposure value can be sensitive to even small changes in the location of the most exposed residence (Graham, 1993: 18).

Finally, the EPA's assumptions about exposure duration and lifestyle were questioned (CONSAD, 1989). The EPA assumed that residents live at the same residence for seventy years and that they spend twenty-four hours a day inhaling outdoor air near their homes. Some people live their entire lives at the same home, but most do not. Even if people move within the same town, they do not necessarily move to another point of maximum exposure. Some people spend most of their time at home, but many do not. Further, people spend more time indoors than outdoors. While pollutants penetrate from outdoor air to indoor air, the ratio usually is not one to one (NRC, 1991).

Although consultants hired by coke producers found no relationship between communitywide rates of cancer mortality and the location of coke plants, the EPA made only minor changes to its assessments (CONSAD, 1989). The agency concluded that people living near coke plants may be incurring significant excess cancer risks. The EPA's findings were printed in the *Federal Register,* were publicized widely at symposiums and by the environmental community, were reprinted in local newspapers where plants operated, and were used by Congressman Henry Waxman (D-Calif.) to dramatize the need for a stronger Clean Air Act.

When Congress revisited the Clean Air Act in 1989 and 1990, the

EPA was criticized for having a weak record in regulating toxic air pollutants. Only seven toxic air pollutants had been regulated (benzene, radionuclides, and inorganic arsenic were the latest additions), and coke oven emissions were not among them. While the agency was convinced that coke ovens should be regulated further, no consensus existed on how stringently to regulate the industry under the 1970 law.

| The 1990 Clean Air Act Amendments

During the 1988 presidential campaign, George Bush pledged to be the "environmental president" by taking steps such as proposing a stronger Clean Air Act. After he was elected, Bush delivered on this promise and worked with Congress to pass the Clean Air Act Amendments of 1990, which were signed into law on November 15. While the president's acid rain proposal garnered the national publicity, the law also contained an ambitious new program to control hazardous air pollutants (the "Air Toxics Title," or Title III), including a specific provision covering coke plants. But passage of the Air Toxics Title was not easy.

In both the House and Senate, a consensus formed quickly around President Bush's proposal to require all industry to reduce hazardous air pollution by implementing "maximum achievable control technology" (MACT) at major plants. Since MACT would usually reduce emissions by 80 to 90 percent, legislators were still faced with what to do about the residual 10 to 20 percent of industrial emissions. President Bush proposed a second phase of rules (after MACT), only if residual health risks were judged by the EPA to be "unreasonable" based on risk assessment and cost-benefit analysis.

Environmentalists, led by attorney David Doniger of the Natural Resources Defense Council, objected to Bush's residual-risk proposal because it was too discretionary and could allow cost considerations to trump public health concerns. Doniger worked with Mickey Leland (D-Tex.) in the House and David Durenberger (R-Minn.) in the Senate on an alternative approach that would compel industry to meet quantitative risk limits if it were not feasible to achieve the ultimate goal of zero emissions. Doniger's approach embraced quantitative risk assessment as a tool for determining whether or not a plant should be shut down, but prohibited any role for cost-benefit analysis in determining how much residual pollution would be permitted.

The final bills passed by the Senate and House differed in how the residual-risk issue was handled. The final Senate bill called for MACT standards to be followed by residual-risk standards that would protect the most exposed person from excess lifetime risks of 1 in 10,000 and ultimately 1 in 1,000,000. (However, the Senate did authorize a National Academy of Sciences study of the EPA's assessment methods and a bipartisan commission to rethink the residual-risk issue). The final House bill embraced MACT but rejected both the Bush and Leland plans. At the insistence of John Dingell (D-Mich.), the House retained the 1970 law's requirement that public health be protected by an "ample margin of safety," which the courts had already interpreted as requiring a minimum degree of health protection (regardless of cost) plus an extra margin of protection (due to scientific uncertainty about risk).

In negotiations between the House and Senate, conferees retained a risk-based threshold of 1 in 1,000,000, but for a different purpose than originally proposed by Leland and Durenberger. The final law requires that residual emission standards be established for those industrial sources that have not reduced cancer risks to less than 1 in 1,000,000 after eight years following the initial MACT standard. Thus, instead of using a "bright line" (for example, 1 in 1,000,000) as a shutdown standard, Congress mandated the number's use as the EPA intended: as a trigger or priority-setting device to determine where additional assessments and standards were necessary, considering feasibility and costs.

| The Coke and Steel Industry's Reaction

The American Iron and Steel Institute (AISI) and the American Coke and Coal Chemicals Institute (ACCCI) were extremely concerned about the proposed Clean Air Act Amendments' potentially adverse impact on coke manufacturing. According to the AISI, the coke industry had been working for years to improve coke operations and reduce emissions. Milton Deaner, president of AISI, explained to a House subcommittee that the industry had already put in place controls that had reduced emissions by more than 90 percent, and that further control of emissions from the coking process would be difficult.

Deaner was referring to the EPA's regulatory framework for meeting Section 109 of the Clean Air Act Amendments of 1970, which required each state to submit an emissions control plan to bring particulate concentrations within its borders into compliance with the health-based national ambient air quality standards (NAAQS). Coke ovens were a significant source of particulate emissions in some "nonattainment" regions of the country, and therefore coke firms had already been required to control their emissions. Firms had looked for ways to improve operating procedures and technology to curb door leaks, lid and offtake leaks, and charging, pushing, and combustion-stack particulate emissions. Also, in 1976 the Occupational Health and Safety Administration (OSHA) had promulgated a standard to limit the emissions of the benzene-soluble fraction of total particulate matter from coke ovens. Specific engineering and work practice controls were also mandated in the standard. Since much progress had already been made, further progress was perceived to be difficult and costly.

Interestingly, the AISI did not focus on persuading Congress that the EPA's risk assessment was wrong. Instead leaders of the integrated steel industry decided that their objective was to avoid risk-based legislation entirely, even if it meant promoting technology-based standards for their own industry. They believed the second phase of regulation, a risk-based approach with arbitrary goals of specified risk levels, would be unreasonable, impractical, and unachievable. Even if it were the risk-based standard was fifteen years away, the steel industry would not be likely to invest in new coke ovens in that time frame. Deaner testified that there was tremendous uncertainty about whether an existing oven could be retrofitted to meet a 10^{-4} standard, and even whether a new battery could meet 10^{-6}. Industry leaders estimated that approximately 15,000 jobs within the coke industry and 10,000 coal workers' jobs might be eliminated if the proposed risk numbers were mandated and enforced (Masciantonio, 1995).

Unlike the AISI's position in the past, when fighting technology-based standards was the goal, the institute recommended "that the legislation dealing with toxic air pollutants be technology based, requiring MACT for control of emissions." Despite the previous claim that the coke industry was "close to the limits of what can be done to reduce emissions," Deaner argued that the MACT standard in the administration's bill was a "reasonable and practical approach and [could] result in major environmental improvements." The AISI then offered Congress advice on how the legislative process could proceed without bumping into confrontation and gridlock: "In my industry we have come a long way, and we have a ways to go, but it isn't done by theoretical calculations. It is done by installing the technology that we know about, and that can be done under the section 112 [current law] or modifications to it" (Deaner, 1989).

Like Deaner in the House hearings, Walter Williams of Bethlehem Steel Corporation urged a Senate subcommittee to control toxic air pollutants through the application of MACT, as contained in the administration's bill. Williams explained that if further steps were needed for controlling residual risks after MACT, AISI would then support controls based on an "unreasonable risk" test, taking into account the availability of technology and cost. The coke industry did not want a hard and fast threshold; instead, it preferred allowing the EPA discretion to decide what was an unreasonable level of residual risk. "To go beyond the MACT level to a risk level of 1 in 10,000 would require the shutdown of 36 to 39 of the 40 coke oven plants," argued Williams (CAAA, 1989b). "A 1 in 1,000,000 risk level would, according to EPA's estimates, result in the shutdown of all coke production facilities in this country. And in fact, the proposed law would probably preclude the construction of any new coke oven batteries to meet that level in the foreseeable future," he warned.

The coke industry recognized that its future was being determined by Congress. Despite the industry's ominous position, others continued to push hard for the risk-based standard. Doniger of the NRDC was encouraged by the industry's willingness to agree that the current law was "broke and needs fixing" and he agreed with them on a technology-based first step. But Doniger believed it to be "extremely important that we retain the public health protection objective that is in existing law as the second phase and [that] it be meaningful and it not be cost benefit based, [but that] it be health based" (Doniger, 1989).

The Industry's "Carve-Out"

The coke industry ultimately persuaded Congress that the proposed risk-based limits would threaten the future of the American coke industry. The merchant coke producers, represented by the ACCCI, splintered from the AISI position and worked with Senator Howell

Heflin (D-Ala.) on a special provision (S. 1630) for the coke industry. The merchant producers were concerned about their ability to meet the initial technology standard (MACT), never mind the residual-risk standard. Merchant producers argued that no bank, knowing that the merchant's operation might fail to meet the future residual-risk standard seven years later, would provide them with capital to invest in the first-tier technology standards. The merchant producers needed twenty to thirty years to amortize their bank loans and to meet the residual-risk provision. The ACCCI convinced Heflin that the MACT and residual-risk standards would shut down the entire merchant coke industry. Heflin's bill (S. 1630), which introduced the idea of an "extension track" for the merchant coke industry, was not passed by the Senate committee but was accepted as an amendment to the final Senate bill (S. 1643).

House-Senate conferees agreed that the special circumstances of the coke production industry required a separate set of MACT and residual-risk provisions (Graham, 1993). Instead of demanding a reduction in cancer risk to 1 in 10,000 within eight years after the MACT standard (the rule that applied to all other industries), the final Clean Air Act Amendments of 1990 offer the coke industry additional compliance options, or "tracks".

One option, the "MACT track," called for existing coke oven batteries to meet MACT by the end of December 1995, and then to meet an "ample margin of safety" standard by December 31, 2003. According to EPA practices, "ample margin of safety" generally means that a cancer risk will be no greater than 1 in 10,000 for a hypothetical individual most exposed to emissions from a plant (Ailor, 1992:7). However, a 1 in 1,000,000 residual-risk number is not uncommon in public health standards. Congress may ultimately change the residual-risk standard based on a recent National Academy of Sciences report and a forthcoming report from the bipartisan Commission on Risk Assessment and Management.

A second option, the "extension track," gives owners of existing batteries an extension to January 1, 2020, for complying with the residual-risk standard. To be eligible for this track, existing coke batteries had to submit to EPA work practice plans by 1993, and they must meet a lowest achievable emission rate (LAER) level by 1998 and a possibly more stringent LAER standard by 2010. As a practical matter, coke firms choosing this track may ultimately avoid risk-based regulation altogether, since the risk limits for the coke industry are so far in the future that the Clean Air Act may be amended again in the meantime, and no one knows what a future Congress might do.

Some firms have chosen a third compliance option that gives coke plants the chance to defer deciding on one of the two tracks until 1998. These firms can "straddle" the two compliance tracks by meeting the requirements of both MACT and LAER. The EPA is expected to clarify the residual-risk standards for the MACT track by 1998, but delays are possible.

If Congress retains a strictly risk-based approach for regulating residual pollution, the final residual-risk number will be critical to this industry. According to the EPA's 1987 risk assessment, all of the operating coke plants exceeded a 1 in 10,000 risk of cancer for the maximally exposed resident. Even a 90 percent reduction in emissions would not necessarily have permitted them to comply with a numerical standard of 1 in 10,000. Some of the plants had estimated risks greater than 1 in 100. If these plants were to cut emissions by 99 percent, they still would not reduce the estimated risk to 1 in 10,000. None of the plants as operated in 1987 had much of a chance to meet a more ambitious 1 in 1,000,000 standard in the forseeable future (Graham, 1993:15).

| Regulatory Negotiation at the EPA

To flesh out the technical details of the special legislative provision applied to coke plants, the EPA formed a "negotiation" group involving representatives from industry, labor unions, environmental and community groups, and state and local enforcement agencies. William Rosenberg, the assistant administrator of the air office at the EPA, chose to negotiate the regulations (as opposed to having the EPA determine them itself) in order to avoid lawsuits and gridlock. The EPA was also led to initiate a regulatory negotiation when its technical analysis suggested that there might be an opportunity for creative "win-win" solutions — stricter standards environmentally than the statutory minimum but also more efficient for industry to implement (Shapiro, 1995).

Over an eighteen-month period, those involved in the negotiation considered many detailed issues, including numerical emissions limits, visible emission monitoring methods, implementation, enforcement, and legal issues. Without this negotiation, during which EPA had an opportunity to learn the "ins and outs" of coke production, it is likely that the resulting standards would have been less stringent (Agnew, 1994). Emission controls were also required sooner than they would have been without the regulatory negotiation (Masciantonio, 1995).

A formal agreement was signed by most of the participants on October 28, 1992. While many suggestions from environmental and community groups were included in the formal agreement, GASP (Group Against Smog and Pollution), a local environmental group with influence in the Pittsburgh area, refused to sign on — but did not challenge the final agreement in court.

Coke oven emissions are difficult to quantify because they often occur through leakage; they are not confined to a duct or stack where they can be easily measured. Consequently, the EPA's negotiation group defined MACT and LAER in terms of the percentage or duration of leaks from doors, lids, and other fugitive emission points. Small leaks were counted the same as large leaks. The resulting MACT regulation restricted, by 1996, the percentage of visibly leaking coke oven doors in a battery to a maximum of 5.5 percent and 6.0 percent for "short" ovens and "tall" ovens (over 16½ feet), respectively. The LAER regulation limits door emissions on existing batteries, by 1998, to no more than 3.8 percent for short ovens and 4.3 percent for tall ovens. Rebuilt coke plants must meet the "super LAER" standard upon startup, which is 2 percent stricter than the regular lowest achievable emission rate. Twelve seconds of visible emissions are allowed for each charge. Compliance is determined daily, based on a thirty-day rolling average. For a new nonrecovery battery that constitutes added capacity at a plant, the regulation is even tougher: the plant is subject to a 0 percent door leakage standard. This standard is considered so stringent that no new by-product plant is expected to achieve it. Added capacity does not always require a 0 percent door leakage rate, however. When a new recovery technology is used, even if capacity is increased, the standard will be determined on a case-by-case basis but will be at least as stringent as LAER.

Through the negotiation process, it was decided that if a firm could prove that it had already committed to rebuilding a battery, or was already in the midst of constructing a new plant, before the CAAA of 1990, its plant could be "grandfathered," meaning that the plant would be subject to the standards applicable to existing plants (to LAER as opposed to super LAER). Three firms (Bethlehem, National, and Koppers) were covered by the grandfather clause and were specifically listed in the regulations.

In addition to emissions standards, the negotiation committee called for specific test methods and coke plant inspection policies. For instance, to determine compliance, each coke battery must be inspected by a person who is trained in a particular inspection method to determine the presence or absence of leaks. The owner or operator of each existing and new coke oven battery is required to develop a written plan that describes emission control practices to be implemented. Further, all plants scheduled to be operating after 1995 are required to install by-pass/bleeder stack flare systems by 1995. These flares, which ignite escaping raw coke oven gas during emergencies, destroy BSOs, volatile organic compounds, and convert hydrogen sulfide into sulfur dioxide.

| The EPA's Benefit-Cost Analysis

The MACT standard was expected to reduce nationwide coke oven emissions from charging and leaks by the end of 1995 by about 80 percent. Emissions from by-pass/bleeder stacks were to be reduced by at least 98 percent. The MACT standards for existing batteries were expected to be achieved without rebuilding batteries, and thus the total nationwide capital cost of MACT for existing batteries was estimated to be only $66 million. Implementation of the LAER standard is expected to reduce nationwide coke oven emissions by 90 percent by the beginning of 1998. After the establishment of the LAER and the installation of flares on by-pass/bleeder stacks, the overall reduction in coke oven emissions is estimated to be 94 percent. Assuming that all batteries will elect to meet the LAER standards, the total nationwide capital cost is estimated to be $510 million, including the costs associated with MACT (EPA, 1993).

The EPA commissioned a cost-benefit analysis by several academic economists. Considine, Davis, and Marakovits (1992b) ran a linear programming model to generate probability distributions for cancer incidence at coke plants. The model was run with and without (base case) the

MACT standards in 1995 and with and without (base case) the LAER standards in 1998. Their findings were that nationwide cancer incidence from coke plants would drop from an estimated 1.1 cases per year in the base case to 0.6 cases under the MACT standards. These health benefits are small compared to the costs. Each avoided cancer case is expected to cost, on average, about $15 million to prevent but could cost anywhere between $10 and $30 million to prevent. Under the LAER standards, the annual cancer incidence would drop only slightly, from an estimated 1.1 cases in 1998 to 0.4 cases. Thus, the cost-effectiveness ratio of LAER would be an average of $30 million dollars per avoided cancer case, with a range from $21 million to $57 million. Although the rules were not found to be particularly cost-effective, Congress had not permitted the agency to base the final rule on cost-benefit considerations, and so this analysis had little influence on the final standard.

| Strategic Options for Coke Producers

Faced with looming regulations and aging ovens, American coke producers had to make strategic business decisions about whether to continue or discontinue their coking business. And for those who chose to persevere, there was the question of which technological and operating changes would lead them to compliance with the Clean Air Act Amendments of 1990. The costs and benefits of these strategies varied for each plant, depending on the age and condition of its ovens, the financial health of the firm, and its access to capital (Considine, 1992a).

Integrated steelmakers are presented with six technology options as they attempt to comply with the 1990 CAAA regulations. (Merchant producers have fewer options, since coke is their only product and their access to capital is very limited.) These six options are summarized in Figure 5.2.

First, producers can choose to operate their existing coke oven batteries, using patchwork solutions as problems arise, like applying luting compounds to door and lid leaks. This strategy can meet MACT requirements but may not meet LAER.

Second, plant owners can refurbish their existing technology by installing new doors and automated door cleaners. In some cases they may choose to rebuild an entire battery from the foundation up.

Rebuilding entails gambling that the high capital costs of coke ovens will be rewarded over the next ten to twenty years with high coke prices. Additionally, pad-up rebuilds are subject to a tighter standard (super LAER) than are existing batteries.

Third, plant owners can choose to shut down their existing plants and adopt alternative or refined cokemaking processes. The nonrecovery or negative pressure oven, for instance, is a proven coking technology that achieves a 0 percent door leakage rate. This technology, a refinement of the old beehive technology, is designed to prevent leakage of all by-products from the oven by keeping the oven pressure lower than the surrounding air pressure. Alternatively, the "jumbo coke oven" is a promising refinement to the traditional by-product coke oven, although it is only in the pilot stage of development. The jumbo coker, used and created in Germany, does create by-products, but compared with the traditional oven's design, the jumbo has many fewer doors and lids per ton of coke produced, consequently lessening the amount of pollution from fugitive emissions.

Fourth, steel producers can reduce their reliance on coke by injecting fuels other than coke into their blast furnace. For instance, pulverized coal injection (PCI), which involves injecting crushed coal directly into the blast furnace, can replace up to 40 percent of the coke currently required in a blast furnace. In 1992, PCI displaced

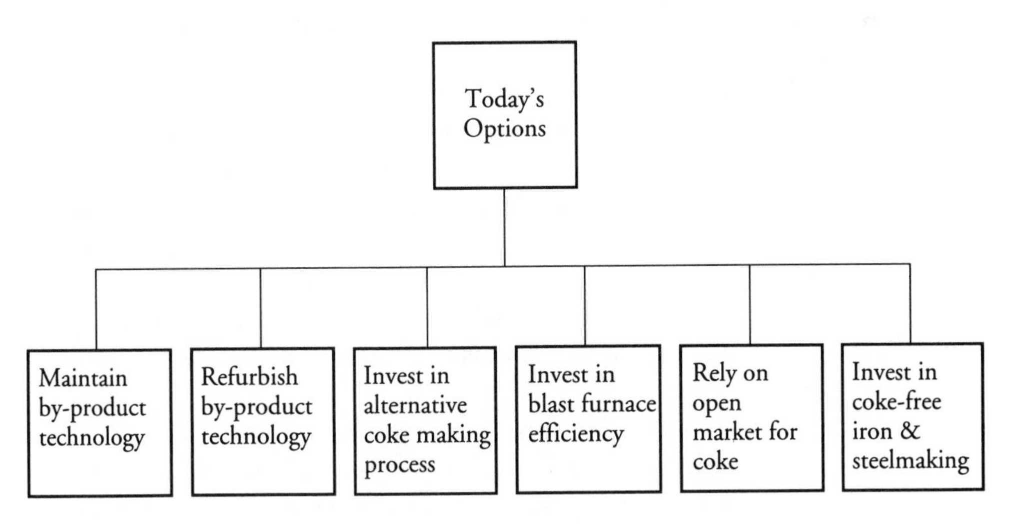

FIGURE 5.2 | Integrated steelmakers' strategic business options.

approximately 3 percent of world coke consumption. According to analysts, this figure may increase by fourfold by 2000 (ITC, 1994).

Fifth, plant owners can choose to shut down their coking operations entirely and purchase coke on the open market from other producers. Although the price of foreign coke fluctuates considerably based on market conditions, this is a viable option for blast furnace owners who require coke.

Finally, integrated steelmakers have the option to invest in scrap-based steelmaking and coke-free ironmaking, and to research direct steelmaking. The electric arc furnace produces steel from scrap recyclables, a process currently penetrating many steel markets that historically have been dominated by coke-based steel. By passing a strong electrical current through the steel, the furnace melts the scrap and allows it to be cast into new steel shapes, bypassing the dirty operations of converting iron ore into steel. With this process there is a higher chance of producing steel that has impurities, because the steel is made from previously used products.

Direct ironmaking and steelmaking involve the production of iron and steel from iron ore and coal alone, thus no coke ovens are required. Besides eliminating most pollution problems associated with coke production and on-going regulatory compliance costs associated with emissions, iron that is produced directly has additional cost advantages over the current processes, including savings from energy efficiency (Unsworth et al., 1989). Also, these direct processes have the potential to produce iron on a smaller scale than typical operations and therefore offer more flexibility in production planning. At present, the large-scale commercial feasibility of direct steelmaking remains uncertain. And even if the initiative proves successful, no one in the American steel industry today foresees this new technology in widespread commercial use within the thirty-year time frame for compliance with the Clean Air Act Amendments of 1990. In addition, even if the process were ready for commercial use now, many companies simply do not have access to the capital needed to invest in it (Deaner, 1990).

The EPA eliminated the prospect of any new by-product plants by setting a door leakage standard of 0 percent for "greenfield" sites and for any existing coke plants where there is an increase in capacity. As has been said, coke oven doors are the single largest source of leaks in the coking process. The 0 percent door leakage standard was designed to promote use of the negative pressure oven, since this is the only cokemaking process known to achieve zero door emissions.

| Preliminary Indications of Firms' Behavior

We surveyed the leading American coke producers (see Figure 5.3) and other industry experts to determine how firms are responding to the regulatory challenge. Although some of the information we sought was difficult to obtain for competitiveness reasons, we were successful in uncovering some clear findings.

Early indications are that several steel and coke producers are making significant emission reductions at existing plants (for example, through door renovations), and there have been some technological investments made to decrease dependence on coke-based steelmaking (three firms surveyed, for example, have installed pulverized injection systems, and one integrated steelmaker is breaking into scrap-based technology). On the other hand, there is little evidence that the EPA's regulatory framework is producing tangible progress toward new and cleaner methods of making coke, iron, and steel. In fact, only one firm surveyed is venturing into an entirely new technology, and there is no interest among American coke producers in the negative pressure oven, despite the intent of the 0 percent door leakage standard to promote the use of this technology. The only firms rebuilding coke batteries are those that were grandfathered and do not have to meet the 0 percent door leakage standard.

There are strong indications that several integrated steel producers are reducing or terminating coke production in favor of purchasing coke on the open market, a trend that has resulted in a growing volume of coke being imported into the United States from Eastern Europe and Asia. In 1992, four firms reported purchasing imported coke (ITC, 1994). This trend will be magnified if several U.S. coke producers are unable to attract the capital necessary to meet the EPA's progressively tighter technology-based controls. A significant decline in the number of operating ovens occurred in 1992 with the shutdown of plants by LTV Steel, Sharon Steel, Bethlehem Steel, and Inland Steel (ITC, 1994). More recently, AK Steel shut down a plant at the end of 1995.

From a worldwide environmental perspective, the reallocation of coke production capacity from the United States to Eastern Europe and Asia is not attractive. Environmental standards tend to be weaker outside the United States, particularly in China, which has emerged as the largest producer of coke in the world.

The experience of Inland Steel Corporation, a large integrated steelmaker, is instructive. In the late 1980s Inland Steel planned to install a negative pressure oven as part of a joint venture with Sun Coal Company. However, the agreement fell through due to financing problems. This was not the first time Inland had tried unsuccessfully to take advantage of new technologies. Inland experienced what other coke producers fear most: huge capital investments poured into a new technology, with disastrous results. In the early 1980s Inland had run into trouble with the first of two new batteries that were built with a pipeline charging system instead of the conventional larry car charging system. This battery was shut down prematurely in 1983. A second pipeline charged battery, started in 1979, never reached expected production capacity. Instead of producing 3,000 tons of coke per day and operating for thirty years (as was expected), this battery produced only 1,800 tons of coke per day and operated for a short thirteen years, shutting down in the middle of 1992. The batteries were demolished

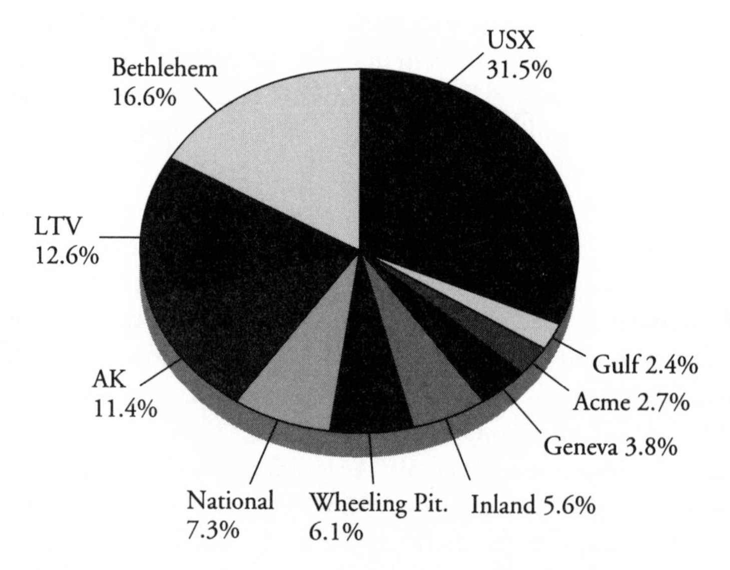

FIGURE 5.3 | U.S. integrated coke producers by market share, 1992.
Source: ITC, 1994, figure 3.3.

and scavenged for usable pieces of equipment. Unwilling to invest more capital into its coking business, Inland Steel made a corporate decision to terminate coke production entirely after learning that it might have to invest $600 million in its by-product coking operation to meet the new clean air requirements. By December 1993, well before the 1995 deadline for the MACT standard, all six of its batteries were shut down. Today Inland purchases coke on the open market from American merchant producers and large integrated firms like USX. Inland also imports a portion of its coke demand from Poland, China, and the Czech Republic.

Shutting down a coking operation is not a cost-free endeavor. Site cleanup, unemployment compensation for workers, and other costs associated with a coke plant closure can be as high as $150 million. A firm also risks weakening its position as an independent and fully integrated steel producer and becomes vulnerable to price swings in the coke market. Several years ago a smaller steel producer, WCI Steel, considered purchasing and renovating an older coke plant to avoid complete dependence on the market supply of coke. The company found that the environmental control costs and potential future liabilities were too great. It is now purchasing coke from a variety of domestic and foreign producers.

Since the Clean Air Act Amendments of 1990, USX Corporation has devoted approximately $150 million to environmental controls at its coke plant in Clairton, Pennsylvania, the largest coke plant in the United States. Despite the costs associated with upgrading its coking operations, USX still believes producing coke is good business. USX has a total of twelve batteries at Clairton, six of which were built before 1956. In 1994 USX announced that it had joined with Nucor Corporation, the leader of "mini-mill" (scrap-based) steelmaking, to apply a new material, iron carbide, and scrap-based steelmaking principles to new product lines. This is one of the first times that leaders of integrated steelmaking and mini-mills have committed to collaborating on new products. The technical details of the joint venture have not been disclosed yet, but the new iron carbide and scrap-based steelmaking is believed to result in fewer air emissions than coke-based steelmaking processes (Holusha, 1994b).

Bethlehem Steel recently finished rebuilding one of its six coke batteries. The rebuilding cost between $150 million and $200 million.

Bethlehem also continues to run an older battery built back in the early 1940s that, equipped with modern environmental controls, meets environmental regulations.

The future of older batteries is nevertheless uncertain, because many rely on unsophisticated environmental controls. For instance, in many cases men working on the ovens are expected to keep an eye out for emissions from leaking doors. When a leak is spotted, a worker then uses a hand tool to clog the emitting crack with a mixture of clay and coke breeze.

In 1992, National Steel Corporation finished replacing one of its three coke batteries in Michigan. The newly rebuilt Great Lakes Division coke battery holds eighty-five ovens and cost the firm more than $300 million. It took two years of construction before the rebuilt battery was ready for operation.

LTV has not yet chosen to rebuild any of the batteries at its three coke plants. Unlike the other firms that have rebuilt coke batteries, if LTV decided to rebuild now, it would be subject to the EPA's super LAER standard. The company has decided to maintain and operate its three existing plants. In addition, LTV announced in 1994 that it plans to team up with two foreign steelmakers to build a $450 million mini-mill.

Starting in May 1989, North American steel producers, with 77 percent of the cost shared by the U.S. Department of Energy (DOE), launched a $30 million pilot plant program to develop direct steelmaking (Dennis, 1991). By March 1994, with more than $50 million expended, direct ironmaking technology was ready for a production-scale demonstration program. However, no U.S. steel producer has chosen to provide the 50 percent cost share required for this next step in the development.

LTV was interested in installing Corex, a German coke-free ironmaking method. The company solicited a grant from the DOE for this effort and was one of five companies to be selected for funding. LTV was awarded $150 million, but since the entire project was estimated to cost $800 million, it was not in a financial position to make a commitment.

| The Slow Pace of Process Innovation

Since passage of the Clean Air Act Amendments of 1990, there has not been a decisive move toward cleaner alternative technologies, such as the negative pressure oven or direct ironmaking. Because new technologies are thought to hold more promise for environmental improvement than simply refurbishing older ovens, why aren't all coke firms interested in investing in alternative technologies? American coke businesses see modifying existing technology or retiring their coke ovens as more economically viable solutions than investing in alternative technologies. Most companies would like to be out of the coke production business altogether, but alternative technologies such as direct iron- or steelmaking require access to capital and high-risk investment.

There are also potential disadvantages to investing in pilot plant research and demonstration projects. A firm making a big investment becomes vulnerable to paying heavily for mistakes — mistakes that other companies can learn and benefit from without making the same investment. At present, the large-scale commercial feasibility of direct steelmaking remains uncertain.

Further, business executives are reticent to move away from a process they know. The blast furnace process is well known and understood. People know how to measure performance when using it, and they fear changing to an unknown method for which methods of measurement have not even been formed. Since firms have substantial capital invested in their blast furnaces, they will not be inclined to scrap this investment absent an extremely attractive alternative process.

The EPA's coke oven emission standards are accelerating the closure of old plants more successfully than they are encouraging the construction of new, state-of-the-art coke batteries. The only firms that can afford to rebuild their ovens are those that had the initial capital to invest in their operations before the EPA's regulatory negotiation. These firms, under a grandfather clause, were permitted to rebuild under more lenient standards (LAER) than the standard that must be met today by firms rebuilding or replacing capacity (super LAER). Greenfields and existing plants adding capacity will be forced to meet unreachable door leakage rates; as described earlier, these batteries are subject to limits based on performance achieved by negative pressure ovens (0 percent door leakage rate). Consequently, it is expected that firms will choose to either shut down their coke batteries in the next several years or lengthen the life of their old batteries, if economical, by applying patchwork solutions to emission problems or until pressured to close by regulatory enforcement.

Regardless of whether an integrated steelmaker makes its own coke or not, it will continue to use its blast furnace technology for decades (or as long as steelmaking is profitable). Consequently, steelmakers will find the coke they need to feed their furnaces. Decreasing U.S. coking capacity has already led to a domestic shortage of coke. The more competitive prices of coke produced in developing countries tempt American blast furnace owners to rely on coke imports.

Stringent environmental regulations in the United States (in conjunction with high labor costs and other factors) have made it difficult for American firms to produce coke as cheaply as it is produced in less regulated countries. Expenditures for control of air pollution at American steel plants totaled $279 million in 1991. Eighty-one percent of that cost was incurred at coke oven sites. The future capital cost of meeting clean air standards for coke ovens alone has been estimated to be as high as $1 billion per year for the next ten years. Therefore the price of foreign coke has become competitive and attracts significant business from American blast furnace owners, who have coke shipped into American deepwater ports, like Baltimore and New Orleans, and then barged as far as the Great Lakes.

In 1992 imports of coke to the United States were modest in magnitude. Of a total of eighteen U.S. firms surveyed by the International Trade Commission, only four steelmaking facilities reported purchasing imported coke. However, compared with other steelmaking countries, imports represented a significant share of domestic coke consumption only in the U.S. market. China, Poland, and the Czech Republic imported no coke, Japan imported only a small quantity, and the United States imported 8 percent of its coke, for a total of 1.9 million metric tons (ITC, 1994).

Up to now, most imported coke has come from developed countries. Approximately 98 percent of the U.S. imported coke in 1992 was from Japan, Australia, and Canada: 80.4 percent, 15.9 percent, and 1.2 percent, respectively (ITC, 1994); only about 0.6 percent (12,000 metric tons) came from China. Preliminary indications are that these trends are changing as Japan cuts back its coke- and steelmaking capacity and developing countries assume a more significant competitive position.

Conclusion

This case study of coke production offers both good and bad news for those seeking to promote risk-based environmental policies. Since it is too early to assess the industry's long-term response to the 1990 Clean Air Act Amendments, some of the predictions and conclusions offered below are necessarily speculative.

One of the myths dispelled by this case study is the notion that quantitative risk assessment is simply a tool used to authorize or legalize industrial pollution. The EPA's risk assessments of coke plants contributed to the developing public consensus that more emission control at coke plants was required. The NRDC in particular was successful in using the EPA's quantitative risk estimates in an advocacy program aimed at curbing hazardous air pollution. While the industry did make some technical objections to the EPA's risk assessments, the industry leadership ultimately lobbied in favor of new technology-based standards in exchange for a twenty-year reprieve from risk-based standards. It appears likely that quantitative risk assessment will play some important role in establishing residual-risk standards.

On the other hand, Congress did not give risk assessment or cost-benefit analysis a significant role in the design of MACT and LAER standards under the 1990 amendments. Even the residual-risk standards are unlikely to provide a role for cost-benefit analysis, unless a new Congress amends the 1990 Clean Air Act Amendments to require cost-benefit considerations.

The reluctance of Congress to incorporate economic analyses into the EPA's rule-making process may have stifled consideration of promising alternatives and may lead to some perverse and unfortunate outcomes. The EPA's analyses raise questions about whether the benefits of the coke production rules justify their rather significant costs. More important, the EPA's rule was designed to encourage investment in negative pressure ovens or new ironmaking technology. Although participants in the regulatory negotiation were acutely aware of the international nature of the coke production industry, preliminary indications are that the EPA's negotiated rule has contributed to a long-term trend of purchasing more coke from other countries rather than producing coke in the United States. The net effect on worldwide emissions from coke production is not likely to be favorable, although residents living near U.S. coke plants will have cleaner air to breathe.

| Sources

Agnew, A., 1994. Personal communication with Amanda Agnew, Environmental Engineer, EPA, December 6.

Ailor, D. C., 1992. *Environmental Issues Facing the Coke Industry in the 1990s* (Washington, D.C.: American Coke and Coal Chemicals Institute).

Air and Waste Management Association, 1995. *Proceedings of an International Specialty Conference,* Pittsburgh, Pennsylvania, April 1995.

Anderson, N. Jr., 1990. "North American Coke Today . . . Red Hot Coke for Red Hot Iron 1990," unpublished manuscript.

Clean Air Act Amendments (CAAA), June 22, 1989a. U.S. House, "Clean Air Act Amendments (Part 3)," *Hearings before the Subcommittee on Health and the Environment, Committee on Energy and Commerce,* 101st Cong., 1st sess.

———, 1990. May 17, 1990. U.S. House, "Clean Air Act Amendments of 1990," *Report of the Committee on Energy and Commerce,* 101st Cong., 2nd sess. (Washington, D.C.: Government Printing Office).

———, September 21, 1989b. U.S. Senate, "Clean Air Act Amendments of 1989," *Hearings before the Subcommittee on Environmental Protection of the Committee on Environment and Public Works,* 101st Cong., 1st sess.

CONSAD Research Corporation (CONSAD), 1989. "Community Health Effects Associated with Coke Oven Emissions," report prepared for USX Corporation, March 20.

Considine, T. J., G. A. Davis, and D. M. Marakovits, 1992a. "Technological Change under Residual Risk Regulation: The Case of Coke Ovens in the U.S. Steel Industry," Department of Mineral Economics, Pennsylvania State University.

———, 1992b. "Costs and Benefits of Coke Oven Emission Controls," Department of Mineral Economics, Pennsylvania State University.

Deaner, M., 1990. Letter from Milton Deaner, President of the American Iron and Steel Institute, to William Reilly, Administrator, EPA, March 19.

Dennis, W. E., 1991. "Lessons from a Decade of Collaborative Research," *Iron & Steelmaker/ISS Publication.*

Doniger, D., 1989. National Clean Air Coalition, "Clean Air Act Amendments of 1989," *Hearings before the Senate Committee on Environment and Public Works,* 101st Cong., 1st sess., September 21, 28–30.

Environmental Protection Agency (EPA), 1984. *Carcinogen Assessment of Coke Oven Emissions, Final Report* (Washington, D.C.: Environmental Protection Agency, February).

———, 1987. *Coke Oven Emissions and Wet-Coal Charged By-Product Coke Oven Batteries: Background Information for Proposed Standards* (Research Triangle Park, N.C.: Environmental Protection Agency).

Farley, J., 1994. Personal communication with John Farley, Honorary Chairman, American Iron and Steel Institute, February 2.

Foster, A., and F. Hojo, 1994. "Coke Battery Automation for Life Extension, Pollution Control, and Industry Survival," ChemTech Consultants, Inc., and Mitsubishi Kasei Corp.

Goldburg, W. 1989. Personal written communication with Walter Goldburg, Group Against Smog and Pollution (GASP), March 4.

———, 1992. *Primer on Coke* (Pittsburgh: Group Against Smog and Pollution, April).

Graham, J. D., 1993. "The Fate of the Maximally Exposed Individual under the 1990 Amendments to the Clean Air Act," Paper presented at the National Research Council, Washington, D.C., 1993.

Graham, J. D., and D. Holtgrave, 1989. "Coke Oven Emissions: A Case Study of 'Technology-Based' Regulation," *Risk: Issues in Health and Science* 1:243–272.

Holusha, J., 1994a. "Steel Rivals Join to Study New Method," *The New York Times,* October 13.

———, 1994b. "Steelmakers' Quest for a Better Way," *The New York Times,* November 30.

Iron Age "Keeping Compliance Costs Down: Metal Producers are Urged to Take Advantage of the Greater Flexibility They are Afforded in Today's Complex Environmental Compliance Standards" (Panel discussion) 9, no. 3 (March 1993).

International Trade Commission (ITC), 1994. *Metallurgical Coke: Baseline Analysis of the U.S. Industry and Imports* (Washington, D.C.: U.S. International Trade Commission).

Lamm, S. H., 1987. Letter from Steven H. Lamm, Consultants in Epidemiology and Occupational Health, Inc., to Dr. Todd Thorslund, EPA, Docket Number A-79-15, August 6.

Lankford, W. T. Jr., N. L. Samways, R. F. Craven, et al., eds., 1985. *The Making, Shaping and Treating of Steel,* 10th ed. (Pittsburgh: Herbrick & Held).

Lloyd, J. W., 1971. "Long-Term Mortality Study of Steelworkers, Part V: Respiratory Cancer in Coke Plant Workers," *Journal of Occupational Medicine* 13:5–68.

Masciantonio, P. X., 1995. Personal communication with Phil X. Masciantonio, former Vice President, U.S. Steel Corporation, March 13.

National Research Council (NRC), 1991. *Human Exposure Assessment for Airborne Pollutants: Advances and Opportunities* (Washington, D.C.: National Academy Press).

Percival, R. V., A. S. Miller, C. H. Schroeder, J. P. Leape, 1992. *Environmental Regulation, Law, Science, and Policy* (Boston: Little, Brown), p. 1354.

Pope, A. C., 1989. "Respiratory Disease Associated with Community Air Pollution and a Steel Mill, Utah Valley," *American Journal of Public Health* 79:623a.

Rosenthal, A., G. M. Gray, and J. D. Graham, 1992. "Legislating Acceptable Cancer Risk from Exposure to Toxic Chemicals," *Ecology Law Quarterly* 19:269–362.

Rueter, F. H., and W. A. Steger, 1990. "Air Toxics and Public Health: Exaggerating Risk and Misdirecting Policy," *Regulation,* Winter, pp. 51–60.

Shapiro, M., 1995. Personal written communication with Michael Shapiro, EPA's Office of Solid Waste, September 5.

Song, Y., 1994. "Modernization of China's Blast Furnaces," presented at the Association of Iron and Steel Engineers Annual Convention, Cleveland. Sept. 1994.

Unsworth, R. E., T. B. Petersen, M. T. Huguenin, and M. J. Reilly, 1989. *Impacts on Integrated Steel Producers Resulting from Regulation of Emissions from Wet-Coal Charged By-Product Coke Oven Batteries* (Cambridge, MA: Industrial Economics, Inc.).

SIX

Coping with Municipal Waste

Alison C. Cullen
Alan Eschenroeder

The municipal waste combustion industry provides a case of significant regulatory and technological change implemented in response to a perceived threat to human health. Single-focus risk analysis catalyzed complex interplays of cultural, economic, and political changes, leading to a somewhat unexpected aftermath. This backlash adds to the richness of the case study, as we shall see in our efforts to unravel cause and effect. It must be acknowledged up front that the management of waste in this country is not merely a practical and technical issue, since for many it encompasses ethical and moral aspects. Certainly a broad spectrum of interested parties, including the public, the regulated industry, environmental groups, and government at all levels, vigorously debate the issue of waste management generally, and municipal waste combustors (MWCs) in particular. Decisions about individual facilities are primarily local, as they have direct consequences for towns and cities; however, long-term nationwide implications grow in significance with the cumulative effect of many local decisions.

In the case of MWCs, health risk assessments (HRAs) preceded several major regulatory events chronologically, without serving as a direct basis for national policy making. Public perceptions, however, shaped by risk assessment language, drove technology-based regulations of MWCs. Differing perspectives on the influence of risk analysis become apparent when we consider who was asking for the analysis and what questions were posed. When site-specific risks are studied for proposed projects, community opposition focuses on exposures that did not exist before the operation of the new facility. The opposition is usually successful in stopping the project. When agencies conduct generic risk assessments to justify policy changes, new technologies are forced into existence by increased stringency. Current concerns are that risk analysis could delay rule making. In this case, the reverse is true, in that risk assessment replaces rule making.

A complete risk analytic framework, such as the one outlined in Chapter 1, was neither applied nor required at the federal level; however, it is interesting to consider the regulation of the MWC industry very briefly in light of the four criteria proposed. First, state-of-the-art science must show that the pollutants of concern are produced or used in the process or activity under consideration. In the case of MWCs, public concern about health risk arose partly as a result of the news that facility emissions contained dioxin.[1] Scientific opinion about the severity and nature of health effects attributable to dioxin has been divided, with many parties revising their views over time.[2] However, two events in the early 1980s, the large government settlement paid to Vietnam War veterans exposed to dioxin-containing Agent Orange and the U.S. government buyout of dioxin-contaminated land in Times Beach, Missouri, achieved permanent notoriety for dioxin, regardless of the shifting sands of scientific hypotheses. Under the second criterion for risk analysis, only processes or industries of high priority warrant regulation. MWCs fit this criterion because they have been subject to federal regulation since the Clean Air Act (CAA) of 1970 and are designated as a potential source category under Section 112 of the Clean Air Act Amendments (CAAA) of 1990. The third criterion implies that MWC regulation and its consequences, which could include shutting down facilities and managing the waste using other strategies, must reduce overall environmental risk. This requirement clearly calls for comparative risk analysis. However, regulation of MWCs did not occur in a broad context of solid waste planning. If it had, a requirement that health risk assessments be performed across the full range of waste disposal alternatives might have resulted in a very different outcome. Finally, since explicit balancing of costs and benefits

[1] In this chapter the term "dioxin" is used to refer to the family of polychlorinated dibenzo-p-dioxins and furans.
[2] More of the dioxin story may be found in Chapter 7.

is not permitted under the Clean Air Act Amendments, this criterion was not considered in the regulation of waste combustors, but economic data (EPA, 1987) offered in support of a rulemaking as cost per cancer case avoided ($10 million to $215 million), along with reductions in the national numbers of cancer cases (two to thirty-one annually) justified the rulemaking. In summary, the development of MWC regulation followed the first two risk-based management rubrics in Chapter 1, but it did not fully follow the last two.

In the first section of this chapter, we trace the post–1970 development of combustion technology for municipal waste disposal. Next, our discussion turns to the origins of health risk assessments of MWCs and their role in the development of state and federal regulation. The third section focuses on the regulatory approaches adopted by two states, California and Connecticut, and their consequent experiences in siting facilities. And finally we describe the evolution of control technology in the MWC industry, with additional technical background provided in the chapter appendix. We conclude with our thoughts about this case relative to the broader theme of the book, the role and impact of risk analysis in environmental protection.

| History and Evolution of the MWC Industry in the United States

Although references to the burning of waste may be found throughout recorded human history, the rise of the MWC industry has generated increasing attention and concern in the past two or three decades, due to several convergent trends. Certainly we have all the ingredients of a pressing waste disposal crisis in the United States. Our population has been identified as the biggest generator of municipal solid waste (MSW), per capita and as a country, in the world. At present, the population of the United States is approximately 250 million, while our rate of municipal waste generation is about 300 million tons per year (Steuteville, 1994). Further, in the past twenty years, thousands of landfills have closed, leaving only about 6,000 operating by 1990, and only about 4,500 operating as of 1993. During the 1980s, new and replacement landfills became increasingly difficult to site, due to concerns about environmental contamination and lack of available land. Meanwhile, landfill tipping fees — the cost of dropping off a ton of

MSW, increased by a factor of two to five or more in densely populated areas, because of increases in the value of real estate and the stringency of environmental regulations. Of the currently operating landfills, 75 percent are projected to close in the next fifteen years (Denison and Ruston, 1990). Finally, the energy crisis of the 1970s prompted a serious look at many alternative forms of power generation. Waste combustion was particularly appealing, since it provided the opportunity to produce power using a guaranteed source of fuel whose perceived value is low. The response in many locations was to draw up plans for the construction of a new MWC; however, these facilities went largely unbuilt (EPA, 1987).

What plans were laid, and what happened during their implementation? As is shown in Figure 6.1A, in 1986 there were 111 operating facilities at which approximately 5 to 10 percent of the MSW in the United States was combusted. At this time, another 210 MSW combustors were in the planning stages. It was anticipated that the new facilities would have startup dates in the mid-1990s, at which point 25 to 35 percent of the MSW stream would be directed to combustion units. Further, it was expected that only a handful of existing facilities would be shut down between the early 1980s and the mid-1990s. The particulate emissions limit[3] for MWCs in the Clean Air Act New Source Performance Standards (NSPS) adopted in 1971 did lead to the shutdown of some MWCs; however, most installed emission controls at that time and continued operation (EPA, 1987). Of the planned facilities, some were built on schedule, but many were canceled or stalled in various planning stages.[4] By 1991 the number of operating MWCs had reached 160 to 170, and the number in planning had dropped to 47. By 1993 the number of units was in the 150 to 160 range. The net result, in spite of the expectation of significant growth in the industry, was only a modest increase in the number of facilities since 1986.

[3] The particulate standard for incinerator emissions adopted in 1971 was 0.08 grains per dry standard cubic foot.

[4] The inventory of MWCs must be viewed as approximate, due to the continually changing status of many of the facilities and the diversity of sources from which they are of necessity derived. Since the various reports cannot be reconciled exactly, we present a range of numbers representing the population of combustors over time.

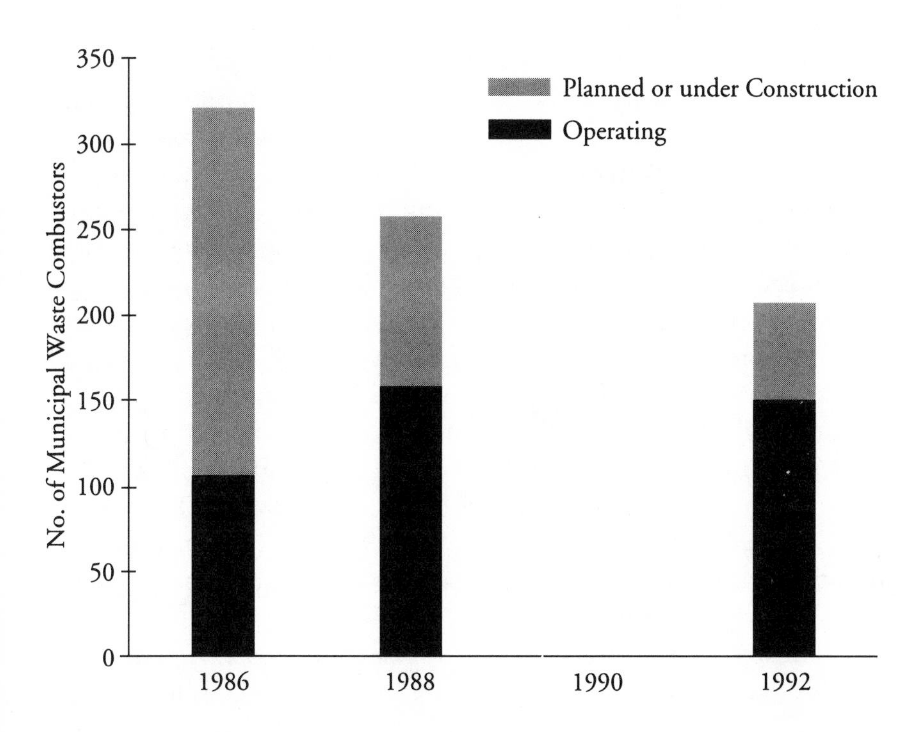

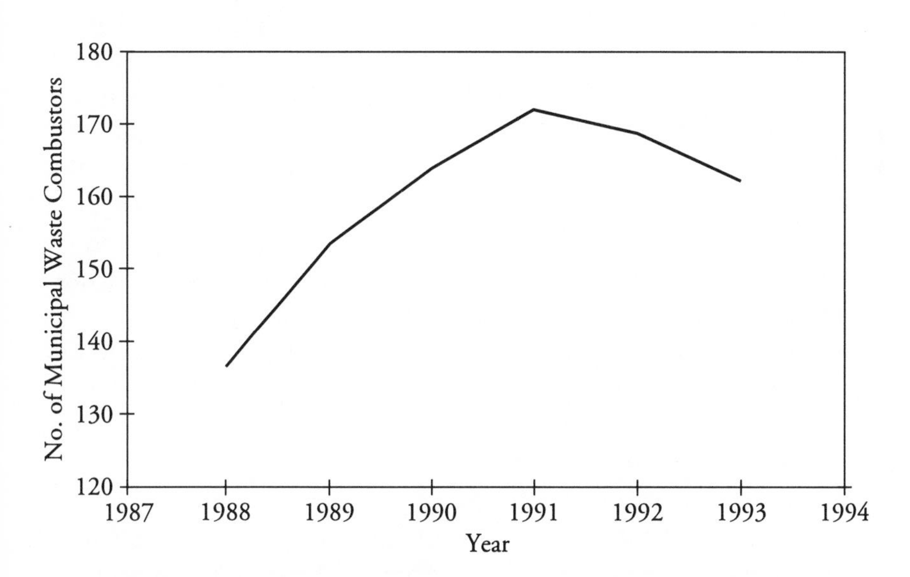

FIGURE 6.1A, B | Municipal solid waste combustors in the United States, 1986–1992.

Sources: Data for figure A adapted from EPA, 1987; Denison and Ruston, 1990; IWSA 1993. Data for figure B adapted from *BioCycle* surveys.

Of course the number of facilities in operation tells only part of the story. The capacity of the operating MWCs is another important measure of the status of the industry. To complicate matters there is inherent uncertainty in the estimation of capacity, and variability in the utilization of that capacity in operating facilities. Although there are some discrepancies in reported capacities, according to data presented by the Environmental Protection Agency (EPA, 1987), the Environmental Defense Fund (Denison and Ruston, 1990), and in the literature (Steuteville, 1994), the total capacity of MWCs in the United States was about 15 million tons per year in 1986 and remained at that level through the late 1980s, accounting for about 5 to 10 percent of the MSW stream (Figure 6.2). By the early to mid-1990s, about 10 to 15 percent of the MSW stream was directed to MWCs (Denison and Ruston, 1990; IWSA, 1993; Steuteville, 1994). The facility population growth tends to underestimate the capacity growth, because the newer units are larger than the older ones.

Putting aside for the moment the lack of success in siting, the degree of geographic concentration of proposed MWCs is notable. Of the MWCs operating as of 1986, most were located in the Northeast, with some in the more densely industrialized and populated sections of the Southeast (Figure 6.3). Many of these areas experienced large rises in land values and a resultant shortage of affordable landfill space as early as the 1970s. Not surprisingly, plans to build new MWCs in areas with rising land values continued to be developed, especially on the East and West Coasts, where the waste disposal issue became quite urgent. For example, the 1986 projections indicated that the biggest increase in capacity would occur in California, where thirty-five new facilities were planned (Figure 6.4). Interestingly, a mere two years later, thirty-two of these had been canceled (see below, "A Comparison of Regulation in California and Connecticut").

MWC Design and Air Pollution Control

There are three basic designs for MWCs used in the United States at present. The most common type is the "mass burn" facility, in which unpreprocessed waste is burned in a chamber through which between 50 and 1,000 tons per day of solid waste may pass. A second common type of facility is the modular combustor, which also accepts unprocessed waste. These MWCs are composed of a series of prefabri-

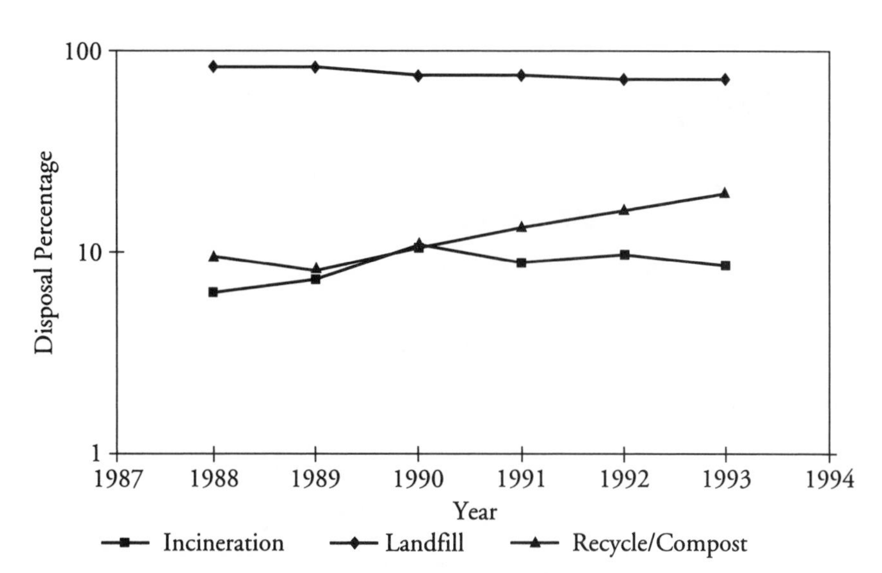

FIGURE 6.2 | Distribution of U.S. municipal solid waste disposal.
Source: *BioCycle* surveys.

cated modular units, each handling between 5 and 100 tons per day. The third combustion design includes a preprocessing step in which the waste is homogenized and formed into a pelletized or shredded "fuel" known as refuse derived fuel (RDF), which is burned in a single chamber. In 1986 the capacity in the United States was divided between these three types of combustors, with 68 percent handled by mass burn combustors, 9 percent handled by modular units, and 23 percent preprocessed at RDF facilities. Over time a slight movement toward RDF facilities has occurred. By 1992, 66 percent of combusted MSW was directed to mass burn facilities, with 4 percent in modular units and 30 percent in RDF facilities (IWSA, 1993).

All three of these designs may be configured for energy recovery — that is, to provide process steam or electrical energy. Such units are classified as waste-to-energy (WTE) combustors. In mass burn and modular systems, combustion gases heat water to produce steam or electricity during or after combustion of the unpreprocessed waste. In an RDF operation, waste is preprocessed to improve the efficiency of the burn for energy production. With only 38 percent of the heating value of bituminous coal and 26 percent of the heating value of distillate fuel oil, unprocessed MSW makes a poor fuel. Moreover, the heating value of the

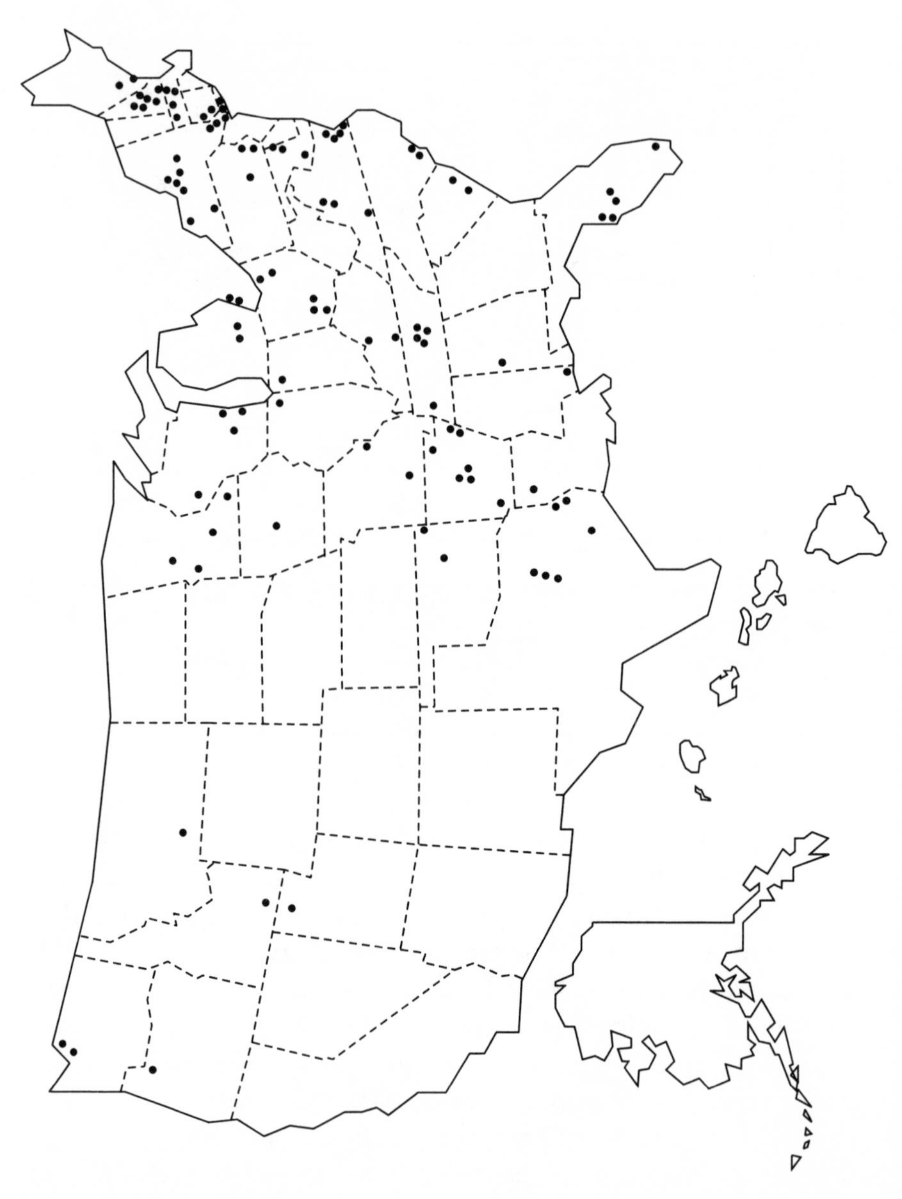

FIGURE 6.3 | Regional distribution of existing municipal waste combustion facilities in 1986.

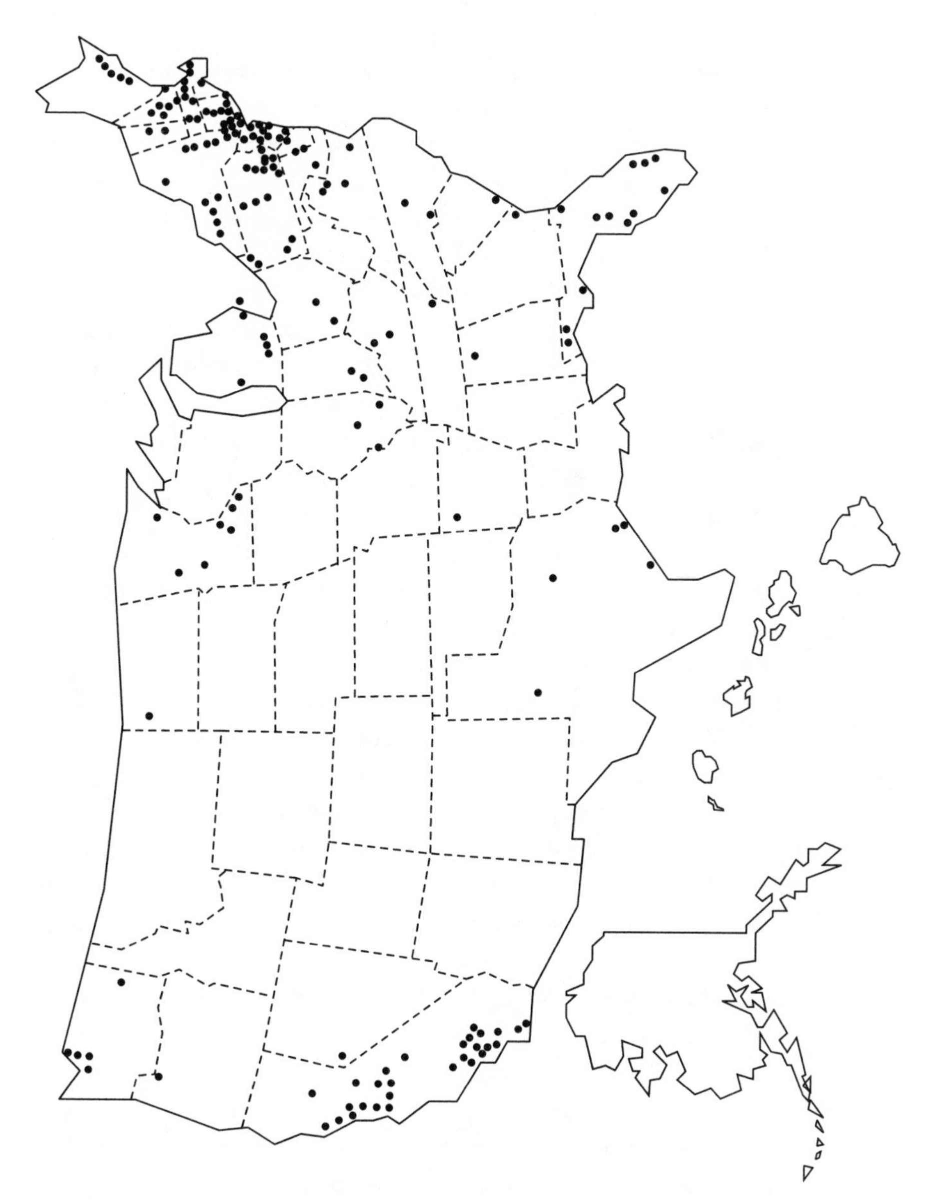

FIGURE 6.4 | Regional distribution of planned municipal waste combustion facilities in 1986.

waste varies widely and unpredictably, due to its heterogeneity. In an RDF operation, bulky or noncombustible items (which compose as much as 45 percent of the MSW stream) are removed and the remaining waste shredded or pelletized for combustion alone or with a supplementary fuel. At present, 83 percent of the MWCs in operation in the United States are designed for WTE capability (IWSA, 1993).

Although the early history of waste burning was marked by the production of dirty, smoky, and often odorous emissions, much progress has been made since the Clean Air Act of 1970 — along with later amendments — to reduce emissions of various contaminants. In state of the art facilities, a series of devices between the combustion chamber and the stack process the flue gas to control the contaminant emissions reaching the atmosphere. The major components targeted for removal from the stack gases include acid gases, such as hydrochloric acid (HCl) and sulfuric acid (H_2SO_4), heavy metals and organic compounds, which may be bound to particulate matter or in a gaseous form, and oxides of nitrogen (NOx).

The history of the development and adoption of various pollution controls by MWCs is documented in Table 6.1. Control technologies designed to neutralize acid gases also limit dioxin emissions. Dioxins form at intermediate temperatures far below those in the combustor. Consequently, faster cooling after combustion forms less dioxin. Spraying in a lime and water mixture, as is used for acid neutralization in modern MWCs, has the added benefits of rapid quenching and solid particle formation; the quenching inhibits dioxin formation, and the particles take up much of the dioxin that does form. The dioxin-bearing particles are then removed from the stream using either electrostatic fields or fabric filters. A more detailed discussion of the technologies employed to clean flue gases is found below, in "The Introduction of Modern Technology and Subsequent Trends," with technical supporting information located in the appendix to this chapter.

Returning to Table 6.1, we see that in recent years there was a massive and rapid increase in the number of facilities operating these pollution control devices. Between 1982 and 1993 the percentage of facilities controlling particulate matter with modern technologies — that is, electrostatic precipitators or fabric filters — increased from 76 percent to 91 percent, with a decided shift from the former technology to the latter. Over the same period, the fraction using dry scrubbers

TABLE 6.1 | Types of air pollution control equipment and percentages of municipal waste combustors incorporating them, 1982–1993

	1982	1984	1986	1988	1991	1993
Electrostatic precipitator	59.0%	52.7%	46.2%	41.3%	36.5%	40.4%
Baghouse or fabric filter	16.9	16.1	34.5	44.3	53.3	50.3
Dry scrubber	6.0	8.0	35.1	43.3	51.8	50.9
Wet scrubber	7.2	4.5	5.3	3.0	4.6	6.4
Afterburn or two-chamber system	19.3	14.3	15.2	12.4	10.7	9.4
NOx control	0.0	0.9	1.8	2.5	14.7	17.5
Mercury control	0.0	0.0	0.0	0.0	0.0	6.4
Other technology	1.2	0.9	4.1	2.5	2.5	1.8
Nothing used	4.8	6.3	4.7	3.0	4.1	2.9
Total number of facilities surveyed	88	113	166	201	202	171

went from 6 percent to 51 percent, while wet scrubber usage remained fairly constant. The use of NOx controls also leapt significantly, from zero in 1982 to 18 percent of facilities in 1993. These great strides in developing and installing air pollution control technology resulted in the much cleaner operation of MWCs.

| How Did Health Risk Assessments of MWCs Originate?

State and Local Agencies Ask for HRAs

Early in the 1980s state and local governments began to require health risk assessments for municipal waste combustors. The impetus came primarily from the prospective neighbors of proposed MWCs, who began to demand information about the potential risks. HRAs were performed exclusively for new facilities, since the assessments were part of the siting process. The risks of alternative disposal facilities, such as landfills, were generally not compared with those of MWCs. The discovery by researchers in the Netherlands, Canada, Japan, and Switzerland of dioxin in the ash released from MWCs raised public concern, since in toxicological experiments, dioxin acted as a potent carcinogen in animals, as well as being acutely toxic. Scientists at the

Dow Chemical Company (Bumb et al., 1980) imputed fire as a general cause for dioxin formation. Their hypothesis was that any combustion process, be it wood burning in a fireplace or gasoline burning in an automobile, would form dioxins. Their qualitative analyses suggested that dioxin emission sources pervaded populated areas. Subsequent research refined the inventory of sources of dioxin and identified MWCs as a high-profile emission source. Given the potent carcinogenicity attributed by regulators to the dioxin mixture emitted, this family of compounds dominates the cancer risk calculations in HRAs for waste combustors (Levin et al., 1991).

Approaches to assessing potential human health risks from MWCs have evolved over time. Although HRAs have taken into consideration increasing numbers of contaminants, exposure routes, and potentially affected individuals, risk estimates have remained in the range of about 10^{-5} to 10^{-7}, with many results clustered around 10^{-6}, or one in 1 million (Levin et al., 1991). There are several reasons for this apparent stability of the risk estimates. For example, improved measurements of the emissions testing have quantified the important individual chemicals emitted from MWC stacks. Knowing the levels of individual compounds, the risk analyst improves estimates of the toxic potential of the emissions. Despite the lengthening list of contaminants included in the calculations, none has a carcinogenic potency even close to that of the dioxin compounds; thus, in many cases the risk estimates, dominated by the dioxins, have been unaffected.

By the mid-1980s, risk assessors began to consider the absorption of contaminants through the skin and digestive tract, not solely through inhalation, as assumed in the earliest HRAs. HRAs began to include estimates of risk resulting from dermal absorption and by ingestion of a range of items from affected areas, such as fruits and vegetables, meat and dairy products, fish, mother's milk, soil and house dust (inadvertently ingested), and water. The early assessments had provided estimated lifetime risks of cancer to a hypothetical individual breathing the most contaminated air expected to be found in the vicinity of an MWC. Moreover, the geographical diversity afforded by air quality models had generalized the results to the entire neighboring population. Also, the earliest analyses had focused on a hypothetical individual exposed for seventy years to the most contaminated environment. During this period of evolving HRA technique, additional pollution controls were added to MWC designs to reduce emissions and projected environmental concentrations. In spite of the greater numbers of contaminants accounted for, and the multiple exposure pathways being considered, the magnitudes of risk estimates remained about constant.

At the same time, scientific understanding of the movement and transport of contaminants through the atmosphere, bodies of water, and the soil was improving. The realization that large populations living at some distance from a facility could be affected as contaminants dispersed motivated risk analysts to examine population risk in HRAs. Population risk results show how many people experience each different level of risk. They vary greatly, depending not only on the level of contamination at each receptor (location), but also on lifestyle factors: particular food preferences, exposures to soil around the home, dietary importance of home grown foods. State-of-the-art HRAs now analyze the roles played both by variability and uncertainty in the estimation of risk.

One of the landmark HRAs was for the Brooklyn Navy Yard incinerator. This facility was to be the first of eight large MWCs planned to relieve the overburdened landfill network of the New York City metropolitan area. An initial HRA, prepared by F. C. Hart Associates (1984), met with criticism from environmentalists (Commoner, Webster, and Shapiro, 1984), who sparked a public debate with their report that health risks due to dioxins had been underestimated. In response to this, the HRA was updated (F. C. Hart Associates, 1984), but the analysis was still restricted to dioxins that were inhaled. A distinguished board of external peer reviewers generally endorsed the updated version and indicated significant weaknesses in the report by Commoner and colleagues. A third, multipathway version of the report appeared four years later (Smith et al., 1988). The HRAs were commissioned by the City of New York Board of Estimate, at the request of the Department of Sanitation.

The tone of press coverage on dioxins, defined during reporting on the Agent Orange case brought by Vietnam veterans and the evacuation of residents in Times Beach, Missouri, fueled public fear in New York about this toxic compound. In other parts of the country similar fears were arising. For example, in California the public's concerns about environmental and health impacts of MWCs were brought to the state legislature, which in turn requested of the California Air Resources Board (ARB) a report detailing impacts on air quality. This began a chain of events leading to the requirement of an HRA as part of every MWC proposal in California.

EPA Studies Risks and Makes Technology-Based Rules

The federal government traditionally regarded solid waste health concerns as local issues. As we will see, the entry of the EPA followed that of state and local entities. The Clean Air Act Amendments (CAAA) of 1977 required the analysis of only the ambient impacts of criteria pollutants; that is, those for which National Ambient Air Quality Standards (NAAQS) had been set. These studies were a required part of a new source review in areas where the NAAQS were violated, and prevention of significant deterioration (PSD) permits were required where the standards were attained. It is a matter of historical record that progress toward the Section 112 (CAAA) regulation of hazardous air pollutants (HAPs) was so slow in the 1980s that the main toxic emissions of MWCs were not addressed by the federal government during that decade. The early applications of Section 112 to other pollutants reveal that HRAs still had not been performed after 1990. Technology-based regulation of HAPs would have been offered to afford an "ample margin of safety" to satisfy the statute. This language allowed legislators to assure their constituents that they were fully protected, the implication being that there was zero risk—although in fact risk levels could not be zero for HAPs that have no health threshold. Believers in absolute safety (zero risk) as a goal were troubled by this dichotomy.

Indirectly, an HRA could be required for the regulation of air emissions through Section 111 of the Clean Air Act Amendments. Section 111 sets New Source Performance Standards and existing source retrofit standards based on the availability and the effectiveness of control technology. This is to be contrasted with the Section 112 philosophy just outlined. If the EPA administrator finds that the installation of control equipment is needed to "protect" health, proposed regulations are promulgated through publication in the *Federal Register,* and adopted (or amended) regulations are placed in Title 40 of the Code of Federal Regulations, probably in Parts 51, 52, and 60. An assessment of potential health effects, an HRA, could be performed, and one often did appear in the technical support documentation for Section 111 actions, along with cost increments per U.S. household.

How did the federal government view the potential health impacts of MWCs? To answer this question we must double back in history and examine earlier HRA activities at the federal level. The multimedia national-scale studies of all of the top-priority water pollutants (in response to the consent decree [8 ERC 2120] in the matter of the *Natural Resources Defense Council et al. v. Train*) as implemented through the Clean Water Act of 1977 probably represented the EPA's first comprehensive national-scale health risk assessment project. Other HRAs on specific regulatory problems began to appear in the EPA and other federal agencies. A panel convened by the National Academy of Sciences in 1981 set out to examine the merits of separating science from policy and to consider the establishment of uniform guidelines for HRAs. In 1983 the academy published the famous "red book" (NRC, 1983), the essence of which was adopted as agency policy by the EPA over the next three years.

The spotlight shifted to MWCs beginning with the congressional mandate in Section 102 of the Hazardous and Solid Waste Amendments of 1984, which directed the EPA to provide a report covering: "(i) the current data and information on the emissions of polychlorinated dibenzo-p-dioxins from resource recovery facilities burning solid waste; (ii) any significant risks to human health posed by these emissions; (iii) operating practices appropriate for controlling these emissions." Evidently this project languished for a couple of years until the Natural Resources Defense Council (NRDC) and the states of Connecticut, Rhode Island, and New York petitioned the EPA on August 5, 1986, to regulate air emissions from MWCs under Sections 111 and 112 of the Clean Air Act Amendments. A nine-volume report (including a national-scale HRA) responding to these points appeared in June 1987, at a time when there were 111 MWC facilities in operation and 210 planned or under construction in the United States. The bulk of these were planned for the Northeast corridor. The HRA focused on the risk resulting only from inhalation of dioxins, but a comparison of risk estimates between existing units and those planned for construction speak volumes about improvements in control technology. The estimate of annual cancer incidence across the U.S. population from the existing units spanned the range from three to thirty-eight cases, and for the projected units from two to twenty-two cases. An assumption that "dry scrubbers combined with high efficiency particle collection devices" would be applied across the industry brought the risk ranges down substantially, to 0.2 to 3 cancer cases for retrofitted existing units and to 0.3 to 1 case for projected units. Thus, the newer technologies reduced risks by more than 90 percent. These results provided a rationale for the formulation of NSPS, although exposure pathways other than inhalation were ignored. Despite this omission, the HRAs illustrate the value of comparative risk analysis. The problem was, new MWCs were compared with old MWCs, but the key comparison of landfills versus MWCs was overlooked. The EPA's Office of Air Quality Planning and Standards had already released guidelines for multiple pathway exposure and risk assessment for MWC emissions (EPA, 1986), but applying the new methodology to scores of facilities would have delayed the nine-volume study substantially.

Beginning in 1971 the NSPS for municipal waste combustors required controls on the emission of suspended particulate matter, based on available technology. By the time the health risk considerations came into play, the NSPS particulate emission limit stood at 0.08 grains per dry standard cubic foot, but no specific toxic pollutant limits were set. The November 15, 1990, amendments to the Clean Air Act included Section 129, which specifically regulates solid waste combustors. This section was the EPA's response to the consent decree in *State of New York et al. v. Reilly* (No. 89-1729 D.D.C.), which set deadlines for the EPA administrator to sign proposed standards for new facilities. Prior to this decree, however, the EPA published an advance notice of proposed rulemaking (ANPRM) on July 7, 1987, to regulate both new and existing MWCs under Section 111 of the Clean Air Act Amendments. This checkered history of rule making was pushed along in fits and starts by litigation by the NRDC, the state of New York, and others, compelling the EPA to comply with acts of Congress as well as its own commitments to regulatory schedules. The new source performance standards finally appeared in 1991 in the form of requirements on good combustion practices, organic emissions (as total dioxins or furans equaling 30 ng/dscm), metal emissions (measured as particulate; 34 mg/dscm), acid gas emissions (SO_2, HCl), nitrogen oxide emissions, and monitoring standards.

With the introduction of the 1991 NSPS and emission guidelines, the EPA was mandating source controls based indirectly on health risk considerations, although not as a direct result of risk assessment. At the time of the rulemaking, the NSPS emission guidelines notice (FR, 1991) stated that the range and uncertainty of potential health risks were so great it was impossible to regulate MWCs under Section 112 of the Clean Air Act. Thus, the 1987 risk assessment was not updated, and the NSPS are based on best demonstrated technology. It should not be overlooked that the wave of public opinion, driven by risk analysis, forced the adoption of the technology.

In September 1994, as a result of a suit by NRDC and the Sierra Club, the emission guidelines of the previous promulgation (the 1991 NSPS) were withdrawn. Newly proposed regulations replaced best demonstrated technology with maximum achievable control technology (MACT) for the regulation of MWC units. These standards require a technology that reflects "the maximum degree of reduction in emissions of designated air pollutants taking into consideration the cost of achieving such emission reduction" (FR, 1994). The new standards are based on the performance of the top 12 percent of operating facilities in terms of pollution control — proof that they are technologically feasible. The choice of the top 12 percent is mandated by the CAAA to set the so-called MACT floor — a level of control that must be equalled or exceeded by the newly installed technology.

| A Comparison of Regulation in California and Connecticut

As a result of both the history of federal regulation of MWCs and the concerns of individual communities, there has been a great deal of local input during the permitting and siting of new facilities. Consequently, there are significant requirements at the state level related to the siting and operation of facilities. Since MWCs are a top-ranked potential source category under Section 112 of the Clean Air Act Amendments of 1990, HRAs will eventually be performed to satisfy federal requirements after adoption of designated technological controls; however, many site-specific HRAs have been performed to fulfill state or local requirements. The residual risk assessments required by the CAAA will follow the technology applications by many years. Even when they are performed, they will be necessarily generic.

The impact of requiring HRAs on the state level is explored below through a comparison between California, in which facility-specific HRAs are required, and Connecticut, in which a generic HRA is used to support an ambient standard. Of course, there are many other potentially important differences between these two states, including political and demographic factors, which may partly explain their individual histories with this type of regulation. Still, the comparison is revealing. The effects of HRAs appear to extend well beyond the production of risk estimates for proposed decisions. The focus of the debate has little to do with the acceptability of the numbers, but rather involves the admission of nonzero risks for this particular choice. The fact that a risk exists seems to eclipse the size of that risk in the public discussions of site-specific HRAs.

Connecticut

In Connecticut the application and permitting process for MWCs does not require that an HRA be performed. So how does the state regulate MWCs? In addition to meeting the federal regulations described above, MWCs are required to demonstrate that they will not cause the state to exceed its ambient air quality standard (AAQS) for dioxins[5] and that they meet the hazardous air pollutant regulations under Section 29 of the state law. In addition, every facility must perform an ambient impact analysis, demonstrating that the NAAQS for criteria pollutants will not be violated by the presence of the facility, even in the location where air concentrations are highest. Compliance with these regulations requires measurement of the concentrations of a range of compounds in the environment and in the MWC emissions. Individual construction permits are written based on the characteristics of each facility, to ensure protection of public health in the state. As of 1995 operating permits are also required for MWCs burning twenty-five or more tons per day. Impact modeling and technology choices govern the permitting decisions.

Although no facility-specific risk assessment is required in Connecticut, risk assessment has been indirectly involved in the siting process through the development of the state's dioxin AAQS (Rao and Brown, 1990). In 1986 Connecticut made history by becoming the

[5] The dioxin standard is based on annual averaging with dioxin specified as 2,3,7,8-TCDD equivalents, using 1987 toxicity equivalents.

first state in the country to adopt an AAQS for dioxins, at 1 pg/m^3. The basis of this standard was a generic but detailed HRA, which accounted for both cancer and reproductive effects, and considered multiple exposure pathways (F. C. Hart Associates, 1987). Promulgation of the standard required a cooperative effort between the two agencies charged with human health protection in Connecticut, the Department of Health, which handles risk assessment issues, and the Department of Environmental Protection, which handles risk management issues. This unique approach to regulating sources of dioxin emissions en masse is intriguing, but a lingering question remains. If the dioxin AAQS were exceeded, how would the facility responsible for this violation be identified? Presumably a source apportionment could be carried out using receptor modeling, a potentially difficult undertaking. Fortuitously, to date the highest concentration of dioxin in ambient air in the state, as measured by the stationary network of monitors, has been 0.037 pg/m^3, well below the standard (Leston et al., 1991).

Returning to a more general question, how has the MWC industry fared in Connecticut? Of the eleven facilities projected in 1986 by the EPA (EPA, 1987), five have come into operation, bringing the total number of operating facilities in Connecticut to seven. The new facilities represent an increase of 5,320 tons per day (tpd) over the 1986 capacity of 485 tpd. The new capacity also constitutes a significant fraction of the 8,520 tpd increment that would have been realized if all eleven projected facilities had come on line. From these numbers it would seem that waste combustion is thriving in Connecticut, while the minimization of impacts on environmental and human health is achieved via a suite of state and federal regulations. However, there are two footnotes to this conclusion. First, it is notable that four of the five new facilities came into operation in 1988. It is not known whether these early sitings reflect the current climate. In fact, recent cases indicate a possible change (Leston, 1994). Recently two wood-burning facilities applied for construction permits and encountered significant public opposition. Site-specific HRAs, required as part of the process, were the focus of public concern. Notably, the wood-burning facilities had similar air pollution control designs to state-of-the-art MWCs: spray dryers, scrubbers, and fabric filters. Ultimately the applications were withdrawn. This outcome appears to be in conflict with the successful siting of MWCs in Connecticut — until one takes a closer look. In the case of the MWCs, there is no requirement for site-specific HRAs; therefore, siting does not revolve around convincing particular individuals in a particular location that they will not sustain adverse health effects due to the facility, a very difficult task. For MWCs the decree that no unacceptable risk will be posed by any facility was made for the entire state on the basis of an umbrella HRA.

California

We now turn to California where HRAs are required as part of the permitting process for each MWC. Let's step back to 1980. In March of that year, the staff of the California Air Resources Board delivered a report in response to the state legislature's request to assess the air quality impact of six proposed MWCs. It concluded that dry scrubbers and baghouse controls were most effective for the control of dioxins, the contaminant of greatest concern. In 1981, Southern California Edison announced its corporate policy to limit its generating capacity to alternative and renewable energy sources. In the face of a skyrocketing population forecast, the California Public Utilities Commission projected electric power shortages. Qualifying facilities were given the right under federal law to sell power to the nearest public utility at a rate set by an "avoided cost" formula influenced by recent oil embargoes as well as soaring construction costs. The incentives for electric power generation are illustrated by the projected annual revenue streams for one proposed facility municipal waste combustor (in northern San Diego County). The proponent estimated $17.5 million from energy sales, $5 million in materials recovery, and $2 million in tipping fees. Applications flooded in.

By 1984 the California Solid Waste Management Board conducted a survey and found that thirty-five waste-to-energy facilities were on the drawing boards in the state and one was already under construction. If completed, the combined facilities would produce an aggregate of over 500 megawatts (MW) of capacity. During that same year the California Department of Health Services (DHS) was circulating a draft policy proposing that carcinogens be regulated as hazardous air contaminants. Simultaneously the Air Resources Board (ARB) expanded the 1981 resource recovery staff report into the famous "pumpkin book" (named for its color) to provide technical guidance for newly proposed resource recovery units. The information in this volume became a nationwide benchmark for the subsequent regulation of MWC emissions. The DHS risk assessment policy, coupled with the ARB pumpkin book, set the stage for permit requirements for every California MWC application. Facilities generating more than 50 MW were licensed by the California Energy Commission (CEC). The HRA portion of each application for certification was reviewed for completeness by the DHS. Questioning about the HRA part of the application dominated the CEC hearings and workshops on each project. Permit applications for facilities smaller than 50 MW threshold were processed by the local Air Pollution Control Districts (APCDs). Each district required a health risk assessment and referred it to the DHS. In one case the CEC attempted to wrest control from the APCD by aggregating several small facilities into a single unit greater than 50 MW, but the courts struck this down. The authority of an APCD to require HRAs and to use an acceptable risk threshold was established under the statutory nuisance provisions of the Health and Safety Code, as a result of the decision in *Western Oil and Gas Association v. Monterey County Unified Air Pollution Control District* (49 Cal. 3rd 408 [1989]).

An early landmark case in California was the HRA of a waste combustor proposed for the northern part of San Diego County in California (NCRRA, 1984). In the original version, the Air Resources Board was responsible for oversight on the portion pertaining to emissions, the San Diego Air Pollution Control District (SDAPCD) was responsible for approving dispersion modeling, and the DHS was responsible for exposure and dose-response assessments. Litigation drove this multipathway HRA into a second edition three years later (HDR Techserv, Inc., 1987) with an expanded list of pollutants and refined exposure factors. Other HRAs followed for regional projects serving the main metropolitan areas; for example, Los Angeles (Eschenroeder et al., 1986), San Francisco (Greenberg et al., 1985) and San Diego (Eschenroeder, Petito, and Wolff, 1987), three of the many examples of HRAs going through the CEC and APCD license or permit processes. Many others went through the APCD new source review process, which required HRAs. But since January 1, 1987, all MWCs seeking air permits have been required by Chapter 1134 California Statutes 1986 to submit HRAs.

So, how has California fared in its quest to site MWCs? The facts are rather stark. Of the original thirty-five facilities planned in 1986, only three have been sited and put into operation, while one operating facility was shut down. The new facilities represent a total of 2,180 tpd of added capacity, in contrast to the 42,522 tpd originally projected (EPA, 1987). This finding leads us to wonder about the fate of more recent applicants who were also required to perform a health risk assessment. It is very difficult to determine the impact of the requirement. Few if any new applications have been filed since the thirty-two facilities failed to survive the planning stages. It seems as if in California the HRA process has lead to public opposition to new facilities — all new facilities, regardless of their particulars. This suggests an indifference to the risk estimates in the HRAs, the numbers expressing the level of risk imposed by a particular facility on particular individuals who are in some way exposed to the emitted contaminants. If risk to health were a deciding factor in siting decisions, then we might expect a lower success rate for facilities for which the imposed risk was higher than those for which it was lower. Yet, history reveals that despite the calculated risk levels, virtually all facilities were equally unsuccessful, an indication that the risk level was not a deciding factor. The contrast in the regulatory approaches of Connecticut and California is in part due to differences in open space availability and in part due to cultural differences between their citizens and their bureaucracies.

| The Introduction of Modern Technology and Subsequent Trends

A number of special features of MWCs make the revolution in control technology truly remarkable. A brief description of the technology follows, while the details of the innovations are presented in the chapter appendix. In short, the burning of garbage, like landfill biodegradation of garbage, produces quantities of acid (primarily sulfuric and hydrochloric). They end up in the combustor's flue gas stream. Depending on temperature and saturation conditions, the acid exists in the vapor phase or as a mist. Another emission control problem readily apparent in the early days was that of particulate matter suspended in the flue gas. Wet scrubbing, slurry spray injection, and powder injection are three methods of acid treatment. The latter two methods are considered dry scrubbing. Wet scrubbing brings an alkaline solution into contact with the gas to neutralize the acid. Slurry spray injection is adjusted to achieve complete evaporation of the water, thereby cooling the flue gas; hence its classification as dry scrubbing. Rapid quenching through the dioxin-formation temperature range inhibits dioxin formation. Finally, direct dry injection of alkaline powders is sometimes used for acid gas control. The revolution in MWC design and operation has brought about massively reduced contaminant emissions (see Appendix).

Consideration of Popular Waste Disposal Alternatives

What happens to all the waste originally slated for combustion in the many planned MWCs? Certainly we are recycling and composting more, but that accounts for less than half of the solid waste generated. The rest is placed in landfills. Landfills are not without risks of their own. Toxic contaminants are known to escape from landfills in the form of gaseous emissions and liquid leachate.[6] Two studies (Eschenroeder et al., 1990a and 1990b) compare the health risks of municipal waste disposal in landfills with those in waste-to-energy combustors. The authors found that, on a ton-for-ton basis, landfills pose somewhat higher health risks than MWCs, whether we compare modern technologies or obsolescent technologies.

The comparison was carried out twice, once for a hypothetical landfill and a hypothetical waste combustor in Southern California, and the second time for hypothetical facilities in eastern Massachusetts. Both comparisons were based on disposal of 1,500 tons per day of waste for thirty years. It was assumed in each case that the facilities would be closed after thirty years; however, long-term health risks resulting from the closed facilities were calculated for an additional forty years. HRAs on each facility were performed for two scenarios. The first scenario assumed an "uncontrolled" facility reminiscent of operations before 1970. The second scenario assumed that pollution controls considered state of the art in the 1980s were installed at each facility. The results for the latter scenario indicated low risks for both of the facilities.

[6] Leachate is water from precipitation or other sources contaminated by chemicals that it picks up as it seeps through the mass of landfilled waste. Leachate becomes a concern if it flows into bodies of ground or surface water.

The landfill risk assessments were subject to the following constraints. Under uncontrolled conditions, the landfill leachate was allowed to flow unimpeded to the groundwater aquifer underneath. In the controlled case, the flow was limited by a triple liner below and a cap above. Small probabilities of failure of these protective shields were based on available measurements of actual failure rates. Data on the levels of contaminants in actual samples of leachate from landfills were used in the calculation. Groundwater dispersion of leachate established levels at receptor wells. Contaminant levels in the gaseous emissions were also based on measurements, and air quality modeling determined receptor ambient concentrations. There were also a series of input data selections for the waste-to-energy facilities. Published values for both contaminant emissions from combustors and the reductions achievable with dry scrubber–fabric filter pollution control systems determined the stack gas concentrations. Risks to a hypothetical maximally exposed receptor by direct and indirect exposure pathways were calculated in each case. Local environmental conditions were used in both cases.

The results are summarized in Table 6.2. According to this analysis, MWCs are 100 to 1,000 times less risky than landfills. The results differ greatly because MWCs burn waste in a closed series of unit processes, whereas landfills decompose waste over extensive open areas challenged by weather, soil instability, and seismic events. Why, then, have the landfills not been displaced with all of the MWC capacity that was proposed in the mid-1980s? As noted above, HRAs were not required of landfills as they were for MWCs. In the absence of comparative data, neighborhoods near proposed sites had to balance an option with a quantitative health risk against one for which no risk information was presented. No effective communication or comparison of the risks of both facilities was provided to the communities. A most important factor to keep in mind is that the practical alternative faced by most communities is the replacement of an old landfill (not subject to a health risk assessment) with a new MWC facility. Perhaps more important than the risk numbers themselves is the fact that modern risk management is effective for either facility.

A risk assessment issue frequently discussed in connection with MWC siting are the high emissions events caused by equipment failure or accidents. Current indirect exposure guidelines (EPA, 1993) call for the use of high emissions values for a conservative fraction of the operating cycle of an MWC in order to represent such events. Likewise, landfill liner or gas collection system failures due to operator error or seismic activity must also be accounted for under these guidelines. The comparative risks presented in Table 6.2 envision normal operation of both kinds of facilities at all times. There is no standard guidance for upset gas emissions or leachate discharge from landfills.

New Developments in Federal Regulations

With such recent releases as EPA administrator Carol Browner's combustion strategy, the newly proposed technology-based regulations for MWCs in the *Federal Register* in September 1994, and the external review draft of the EPA's ongoing reassessment of 2,3,7,8-tetrachlorodibenzo-p-dioxin, it is clear that the federal government seems poised to take more definitive stances in debates both about combustors and dioxin. The ultimate impact of these developments is unknown; however, perhaps they portend stricter scrutiny by the EPA, possibly leading to increased confidence in the effectiveness and efficiency of combustion technologies. For the moment, little progress is expected on the siting front, however. Indeed, Browner's 1993 Hazardous Waste (HW) Reduction and Combustion Strategy acknowledges that capacity will be temporarily "frozen" while a national waste strategy is developed in accordance with its stated goals.

Although strictly applying only to hazardous waste, the HW Reduction and Combustion Strategy announced by Browner May 18, 1993, contains language that hints at changes that will reach well beyond hazardous waste. Browner announced her intention to overhaul federal rules governing waste combustion.[7] Previous federal regu-

[7] This process was ongoing at the time of this writing.

TABLE 6.2 | Comparison of alternative methods of municipal waste disposal: lifetime incremental cancer risk for multipathway exposures

	CALIFORNIA LANDFILL	CALIFORNIA MWC	MASSACHUSETTS LANDFILL	MASSACHUSETTS MWC
Controlled	9×10^{-7}	8×10^{-9}	7×10^{-7}	3×10^{-9}
Uncontrolled	4×10^{-2}	5×10^{-5}	8×10^{-2}	1×10^{-5}

lations did not require HRAs for any category of MWCs; however, the HW Combustion Strategy indicates that complete HRAs, including a consideration of indirect exposure pathways, will be part of the permitting process in future. The current oversupply of hazardous waste disposal capacity may render this issue moot for new facilities.

With the final fate of MWCs under the new Clean Air Act Amendments yet to be decided, it is nevertheless clear that MWC regulations have become more stringent but remain achievable technologically. The newly proposed MWC regulations are intended to reach beyond the requirements of the NSPS of 1991. As promulgated, they tighten emissions standards for dioxin and metals, although no HRA requirement is planned. The next frontier is the efficient operation of facilities and equipment to ensure the cleanest burn achievable. Monitoring and inspection requirements provide an avenue for improving operations, since the current gaps between testing allow emission excursions beyond the intended limits to go undetected.

At the same time that new regulations are being developed for MWCs, the EPA has put tremendous effort into improving the information base about dioxin, the most notorious compound in MWC emissions. In June 1994 the results of the EPA's third reassessment of dioxin were released in the form of two external review drafts, the latest drafts in a process whose complexity has thwarted many deadlines. The drafts include a health assessment document for 2,3,7,8-TCDD and a methodology for estimating exposure to dioxinlike compounds. The EPA undertook the reassessment in response to emerging scientific knowledge about the human and environmental impacts of dioxin. New information about the toxicity, carcinogenicity, and other health end points, and about the pathways of human exposure, had come to light since the last reassessment. Some originally believed that this third reassessment would result in a reduction in the cancer potency factor for dioxin and thus loosen regulations in a broad array of industrial applications. It has instead pointed to reproductive, immunological, and other systemic effects as potentially more sensitive end points than the carcinogenic effects previously responsible for driving many risk assessments. Further, in the draft guidance for assessing total exposure to dioxin, indirect pathways of exposure — dermal and ingestion routes into the body — were singled out for comment. Long noted by serious exposure assessors as more significant than the inhalation pathway in terms of their contribution to total exposure, it is now widely held that the contribution of indirect pathways is greater by several orders of magnitude.

| Conclusions

What was the impact of risk analysis in the regulation of the municipal waste combustion industry? Did the use of health risk assessment results strengthen or weaken environmental protection? The answers to these questions are complex and sometimes contradictory. First, as mentioned earlier, HRAs are not required of MWCs by the federal government, although many have been performed to satisfy local requirements. Thus, the risk estimates developed by the federal government were not under any statutory mandate and could not serve as the sole basis of regulation; however, the generic risk assessments did motivate regulatory attention in general. HRAs have clearly and systematically reported that dioxin and other toxic compounds are present in MWC emissions and that disposal of waste by combustion is not a risk-free enterprise. Public reaction to the HRAs has been characterized by qualitative fears and concerns. In California, where site-specific risk assessments are required, facility siting has been uniformly unsuccessful, seemingly unrelated to the level of risk imposed by a particular facility. Interestingly, in Connecticut, where a risk-based state standard for the concentration of dioxin compounds in air stands in the stead of facility-specific HRAs, the public has accepted new sitings. However, for other combustors (for example, wood burners) HRAs are required in Connecticut and have been observed to raise public concern and opposition similar to that encountered in California. Facility-specific HRAs seem to raise general public concern about potential health effects associated with industrial processes or facilities, without clearly communicating or interpreting the magnitude of the risk estimate relative to that of other disposal options.

Current risk-based regulatory approaches require a health risk assessment for a proposed project and compare the results with some selected ("bright line") level deemed acceptable. This protocol is operative whether the HRA is site specific or generic, as in the case of setting an ambient standard or a new source performance standard. The classic permit-compliance approach is embodied in the command/control philosophy. It is simple, but it does not work well for decisions in the face of real uncertainty or competing factors. For example, the agency guidelines control the estimation of risk and thereby may drive results and decisions. As former EPA administrator William Ruckelshaus once said, "A risk assessment is like a prisoner of war, if you torture it enough you can get it to say anything you want."

The environmental review process is an alternative to the command/control process. This approach better insulates the risk assessment from agency control over physical assumptions and content (Broiles, 1988). Its prototype already exists in the federal statutes. That alternative is embodied in the National Environmental Policy Act (NEPA) of 1969 (P.L. 91-190, 42 U.S.C. 4321 et seq.) and is repeated in numerous state laws patterned after it. The NEPA process is to require full and simultaneous disclosure to the public and to the decision makers of all of the critical environmental factors surrounding a decision. The disclosure must describe the present setting, the impacts of the proposed action, and the adverse impacts of the proposed action that cannot be avoided. Applicable guidelines also call for comprehensive assessments of alternatives at the same level of detail as those devoted to the proposed action (40CFR1500.8[a][4]). Importantly, the actual decision is separate from the impact statement. Decisions under this scheme balance social, economic, political, and other factors, along with risk. Further, the process envisioned encompasses the entire waste management system instead of a single facility or a single disposal alternative. On the one hand, public concern has had a positive effect in contributing to industry motivation to develop cleaner and more efficient designs and operations for MWCs to greatly reduce contaminant emissions, and in alerting policy makers of the need for regulation to ensure safety. On the other hand, public concern has remained unassuaged by these technological and regulatory measures, and the siting of MWCs remains virtually impossible in much of the United States, even while the demand for waste disposal options has become urgent.

It is true that great strides have been made to divert recyclables from the waste stream; however, even under optimistic projections the remaining quantity of solid waste is now greater than our *total* waste stream of twenty years ago. The majority of the waste generated in the United States ends up in landfills, which impose less well publicized risks on the population. In our opinion, the risk per ton of waste disposed is greater for landfills than for MWCs. As landfills continue to reach capacity and close, new landfills perhaps will face similar public opposition to that experienced by MWCs, due to the advent of comparative risk assessment in public decision making. The current diversion of waste streams from failed MWC projects into expanded landfills has a limit. Land use constraints and stricter regulations suggest a new era of MWC development. Public scrutiny will demand stricter monitoring and quality control of MWC operations. These factors will certainly govern future MWCs through at least the permit conditions and perhaps through future federal operational standards. Such standards have a long-standing precedent in regulating wastewater treatment plant operations and personnel.

In summary, it appears in this case that risk analysis has helped us win the battle for environmental protection, but lose the war. The systematic revelation of information through risk analysis, the assessment of contaminant emissions and potential human exposures and health impacts, helped to bring about tighter regulation, albeit through technology-based standards, and prompted technological innovation for cleaner MWCs. Ironically it has subsequently become increasingly difficult to site these carefully designed and regulated facilities. This seemingly paradoxical situation has a totally logical explanation. The ultimate effect of the risk analysis depends strongly on who is asking the questions and what questions are asked. If state or local agencies ask for a site-specific risk analysis, and the analysis asks how many additional cancer cases there will be, the community will answer, "We don't want any additional cancer cases — scrap the project." If a regulatory agency, say the Connecticut Department of Health, asks for a generic risk analysis on MWCs, no one would oppose setting an ambient standard that was protective of human health. Meanwhile municipal waste is managed using alternative disposal methods accompanied by risks that go largely unexplored.

| Sources

Broiles, S., 1988. "A Suggested Approach to Overcome California's Inability to Permit Urban Resource Recovery Facilities," *Risk Analysis* 8(3):357–366.

Bumb, R. R., W. B. Crummett, S. S. Cutie, et al., 1980. "Trace Chemistries of Fire: A Source of Chlorinated Dioxins," *Science* 210(4468):385–390.

Commoner, B., T. Webster, and K. Shapiro, 1984. *Comments on the Brooklyn Navy Yard Risk Assessment,* Center for the Biology of Natural Systems, Queens College, City University of New York, Flushing, NY.

Denison, R. A., and J. Ruston, 1990. *Recycling and Incineration* (Washington, D.C.: Island Press).

Environmental Protection Agency (EPA), 1986. *Municipal Waste Combustor Guidance on Exposure Assessment.*

———, 1987. *Municipal Waste Combustion Study,* Office of Solid Waste and Emergency Response, Office of Air and Radiation, Office of Research and Development, EPA/530-SW-87-021 (Washington, D.C.: USEPA, June).

———, 1990. *Methodology for Assessing Health Risks Associated with Indirect Exposure to Combustor Emissions,* Office of Health and Environmental Assessment EPA/600/6-90/003 (Washington, D.C.: USEPA).

———, 1993. *Draft: Addendum to the Methodology for Assessing Health Risks Associated with Indirect Exposure to Combustor Emissions,* Office of Research and Development, EPA/600/AP-93/003 (Washington, D.C.: USEPA).

Eschenroeder, A. Q., P. Guldberg, J. Hahn, et al., 1985. *An Analysis of Health Risks from the Irwindale Resource Recovery Facility,* prepared for Pacific Waste Management Corporation. Pasadena, CA.

Eschenroeder, A. Q., C. Petito, and S. Wolff, 1987. *A Health Risk Assessment for the Proposed SANDER facility in San Diego, California,* prepared for Signal Environmental Systems, March. San Diego, CA.

Eschenroeder, A. Q., S. Wolff, A. Taylor, and D. Burmaster, 1990a: "Health Risks of Alternative Methods of Municipal Solid Waste Disposal: A California Comparison," paper 90-182.3, prepared for the Air and Waste Management Association 83rd Annual Meeting, Pittsburgh, June 24–29.

———, 1990b. "Health Risks of Alternative Methods of Municipal Solid Waste Disposal: A Massachusetts Comparison," Paper prepared for the Society for Risk Analysis, 1990 Annual Meeting, New Orleans, October 7–10.

Federal Register (FR), 1991. "Standards of Performance for New Stationary Sources; Municipal Waste Combustors," *Federal Register* 56, no. 28 (February 11):5489.

———, 1994. "Standards of Performance for New Stationary Sources: Municipal Waste Combustors," *Federal Register* 59, no. 181 (September 20):48200.

Gough, M., 1986. *Dioxin, Agent Orange — The Facts* (New York: Plenum Press).

Greenberg, A. J., and Systems Applications, Inc., 1985. *Supplemental Environmental Information Health Risk Assessment,* Appendix J, Application for Certification to the California Energy Commission, Bay Area Resource Recovery Facility Project, submitted by Combustion Engineering, Inc. Windsor, CT.

F. C. Hart Associates, 1984. *Assessment of Potential Public Health Impacts Associated with Predicted Emissions of Polychlorinated Dibenzo-Dioxins and Polychlorinated Dibenzo-Furans from the Brooklyn Navy Yard Resource Recovery Facility,* prepared for the New York City Department of Sanitation (Draft and Final versions).

———, 1987. *Multiple Pathway Human Exposure and Health Risk Assessment of*

*Polychlorinated Dibenzo-*p*-Dioxins and Polychlorinated Dibenzofurans from Municipal Solid Waste Incinerators* prepared for State of Connecticut, Department of Health Services, February.

HDR Techserv, Inc., 1987. *Risk Assessment for Trace Elements and Organic Emissions* North County Resource Recovery Associates. San Diego, CA.

Integrated Waste Services Association (IWSA), 1993. *Municipal Waste Combustion Directory: 1993 Update of U.S. Plants.* Washington, D.C.

Leston, A., 1994. Personal communication.

Leston, A., J. Catalano, K. Crossman, et al., 1991. *The Connecticut Department of Environmental Protection's Evaluation of the Pre/Post Operational Dioxin Monitoring Conducted at Four Resource Recovery Facilities,* Connecticut Department of Environmental Protection, Hartford.

Levin, A., D. B. Fratt, A. Leonard, et al., 1991. "Comparative Analysis of Health Risk Assessments for Municipal Waste Combustors," *Journal of the Air Waste Management Association* 41(1):20–31.

North County Resource Recovery Associates (NCRRA), 1984. *Risk Assessment for Trace Elements and Organic Emissions,* North County Resource Recovery Associates. San Diego, CA.

National Research Council (NRC), 1983. *Risk Assessment in the Federal Government: Managing the Process* (Washington, D.C.: National Academy Press).

Rao, H., and D. Brown, 1990. "Connecticut's Dioxin Ambient Air Quality Standard," *Risk Analysis* 10(4):597–603.

Smith, A. H., M. T. Smith, R. Wood, and H. Goeden, 1988. *Health Risk Assessment for the Brooklyn Navy Yard Resource Recovery Facility,* Health Risk Associates. Berkeley, CA.

Steuteville, R., 1994. "1994 National Survey: The State of Garbage in America, Part I," *Biocycle* April: 46–52.

| Appendix: Evolution of MWC Control Technology

Wet scrubbers, in the forms of spray tray towers, mechanically aided sprays, or venturis coupled with mist eliminators, have been used to control emissions from municiple waste incinerators. However, they suffered corrosion problems and were not efficient enough as cleanup devices to meet modern regulations. Wet scrubbers are now largely supplanted by dry scrubbers. Dry scrubber systems on MWCs were not new at their time of introduction to the U.S. MWC industry in the early 1980s. They had been installed by Ciba-Geigy in Basel, Switzerland, in the 1970s, by Deutsche-Babcock-Anlagen throughout Germany, and by the Teller Environmental Systems organization throughout Japan. Moreover, this technology had been a subject of exploratory study for utility boiler flue gas desulfurization during the same era by the Electric Power Research Institute. Of all the methods of dry scrubbing, the lime slurry spray dominates in current U.S. practice.

How do dioxins form in flue gas flows? No matter how high the combustor temperature or how long the furnace residence time, dioxin emissions form in the intermediate temperature ranges (say, from 300° C to 800° C) that occur downstream of the combustion chamber, between the heat recovery sections and the stack breeching of the unit. This has been shown in operating MWCs and in laboratory experiments using gas flows through beds of fly ash particulates. Current theory favors a precursor pathway involving chlorinated phenol formation from benzene rings and chlorine atoms, both of which are plentiful in the postcombustion mix. These precursor compounds are adsorbed on the surfaces of fly ash, where they are oxidized and further chlorinated to form dioxins. They then split into vapor and particulate fractions, with the former dominating at flue gas temperatures.

How does dry scrubbing control dioxins as well as acid gases? Rapid cooling in the presence of high particulate contact area is the key to dioxin control in the dry scrubber. The faster the cooling and the lower the final temperature, the greater the removal (or, more accurately, the prevention) of dioxin. This capability of the dry scrubber system came almost as a bonus with the acid gas control for which it was originally intended. Although some of the dioxin is captured in the residual scrubber particulate, a portion of it resides in the vapor phase at this point; therefore, the flow leaving the scrubber must undergo particulate and vapor-phase organic pollutant removal. With the exception of mercury, the metallic pollutants are mostly in the particulate phase.

"Downstream" from the scrubbers is the particle removal stage of the process. The lime slurry dry scrubbing process adds significant particulate loading to that already produced in the combustor: namely, lime residue and the salts from acid neutralization. Combustor particulate is composed of both soot (which is loaded with highly conjugated six-carbon rings) and refractory ash particles (inorganic oxides, chlorides, sulfates, and nitrates). Earlier MWC designs relied on multiclones or venturi scrubbers. Multiclones use centrifugal force to throw particles out to a collector wall in a whirling flow. Venturi scrubbers use the accelerated flow in a convergent-divergent duct to atomize water droplets by flow shear stresses. The fine droplets entrain particles and carry them off in a liquid suspension.

Modern particle removal equipment falls into two main categories: electrostatic precipitators (ESPs) and fabric filters, which are also called baghouses. The ESP uses a corona discharge in the flue gas flow to induce electric charges on the particles. Following this, the forces set up by intense electric fields between large parallel plates sweep the particles out of the flow and onto the surfaces. Rappers shake the particles loose, and they fall to a hopper at the bottom of the duct. The baghouse is like a giant set of vacuum cleaner bags in a steel framework. Gas is forced to pass through the bag fabric, and a cake of particles builds up on the upstream surface. Reverse-flow or pulse-jet action periodically removes excessive buildup of the caked particles and collects them in a hopper. In the baghouse, the porous cake detains the flow, and the solid surface exposed to the gas flow is much larger than that in the ESP. Fabric stability demands that the gas flow through the baghouse be well below the 300° C dioxin-formation cutoff, but a much higher temperature is tolerated by the ESP. It is important to avoid cooling the flue gas below the condensation point to avoid condensation of acid, which eats the fibers.

Thus, the engineering limits on baghouse operation require a narrow temperature window in the vicinity of 150° C to 250° C to avoid burnout at the high end and corrosion due to condensation at the low end. As the gas flows through the pores, much of the vapor-phase dioxin adsorbs on the cake surfaces. In these processes lie the main advantages of baghouses over ESPs in removing both dioxins and fine (inhalable) particles. This all comes at a price, of course, in space requirements at the plant site and induced draft fan power needed to overcome the pressure drop over the filter cake.

By 1984, the California Air Resources Board's "pumpkin book" made it clear to would-be developers of MWCs that dry scrubbers and fabric filter particulate removal were the control technologies of choice. This same requirement found its way into federal regulations by 1991 via Section 129 of the Clean Air Act Amendments and the ensuing New Source Performance Standards for MWCs. This control equipment, coupled with improved combustor design, embodies the response of the industry. The previously popular dry scrubber and ESP technology is not likely to meet either the California requirements or the NSPS requirements. It can be argued that enough fields (stages) in the precipitator can achieve performance equal to that of the fabric filter, but the cost and space requirements may be prohibitive.

As with any piece of equipment, careful maintenance is needed to assure design performance of the baghouse. Torn or burned out bags can lead to increases in emissions up to a thousandfold. Monitoring requirements, inspections, and record keeping are usually in the permit conditions. To account for suboptimal equipment operations in a health risk assessment, the California Air Resources Board recommends increasing the emission values reported in the source testing database by tenfold during periods of poor operating conditions. It further recommends that the HRA assume that these poor conditions occur 20 percent of the time in the case of organic emissions and 5 percent of the time for metallic emissions. Upsets or emergencies are characterized by uncontrolled emissions for an hour (for example, if 99 percent control is expected, the emission test data are multiplied by 100). The frequency of emergencies is not specified in the ARB recommendations, and MWC operators claim that the unit would be shut down in minutes following an upset. The ARB recommendations regarding off-design emissions have been incorporated in the EPA's indirect exposure guidelines (EPA, 1990; EPA, 1993).

A significant area of development is the recovery of resources: energy for electric power or steam production and materials applications for ash. The old units generally had refractory walls, but the new ones have water walls designed to capture energy. Water walls are furnace walls lined with vertical tubes or passages through which boiler feedwater circulates. Refuse-fired steam generators have all of the equipment found in large fossil-fueled central stations, such as reheaters, feedwater heaters, superheaters, and economizers. These devices maximize utilization of the energy from the burning fuel in a thermodynamic cycle. Many schemes have been devised for the recycling of MWC ash. Every year a two-day international conference is held on this subject. Most of the applications involve ash as an aggregate for structural materials such as concrete.

Health risk assessments have also changed the MWC industry by forcing improved combustor designs. Old MWCs used either a rotary kiln or a traveling grate stoker similar to those in small utility boilers designed in the midtwentieth century, while modern waste-burning furnaces have sloped rocking grates. Two notable proprietary combustion systems are the sloped rocking grate furnaces designed by Martin Gmbh and Von Roll, both German firms. With these designs, which are used on nearly all modern units, each transverse grate element has a fixed axis of rotation. The collective motion of the elements, each rocking or rotating about its axis, moves the burning fuel bed down the slope. The motion is controlled so that completion of combustion occurs at the end of the slope, where bottom ash is kicked off the grate by the last element. The ash falls into a quench pit, where it is collected beneath the firebox. Each grate element is perforated with air passages so that the rotary or rocking motion partially modulates the gas flow rate through the furnace.

In combination with improved combustor design, the automation of combustion management has been another technological innovation for the industry. In old units, the lack of operational flexibility led to poor firing practices and their intermittently excessive emissions of carbon monoxide, soot, and dioxins. In current designs, underfire air, overfire air, firing rate, and grate motion are adjusted to maintain preset gas temperature and excess oxygen programs throughout the furnace, consistent with the desired steam production rates. Continuous sensors feed signals into a centralized microprocessor system, which regulates air flow and grate motion. Refuse is notoriously nonuniform with respect to chemical composition and heating value; therefore, these furnace control systems must be more responsive than those on the average fossil-fuel utility boiler.

SEVEN

Producing Paper without Dioxin Pollution

Kimberly M. Thompson
John D. Graham

On March 15, 1983, Ted Koppel stared out at millions of television viewers watching *ABC News Nightline* and asked a question that the nation had struggled with for over a decade: "How dangerous is dioxin?" Indeed, it had been almost a decade since dioxin achieved notoriety as "by far the most toxic compound known to mankind" in testimony before a 1974 U.S. Senate subcommittee by Dr. Diane Courtney, then chief of the toxic effects branch at the Environmental Protection Agency's National Environmental Research Center (NYT, 1974).

The term "dioxins" refers to the class of seventy-five structurally related compounds formally called polychlorinated dibenzo-p-dioxins (PCDDs). PCDDs and a similar class of compounds, called polychlorinated dibenzofurans (PCDFs, or furans), include two benzene rings connected by oxygen bridges, and one to eight substituted chlorine atoms. The most toxic and notorious of the dioxins is 2,3,7,8-tetrachlorodibenzo-p-dioxin (2,3,7,8-TCDD), which includes four chlorine atoms (hence, tetra) substituted in the 2,3,7 and 8 positions. The term "dioxin" has frequently been used to refer to 2,3,7,8-TCDD only, and this chapter follows that convention. Similarly, 2,3,7,8-tetrachlorodibenzofuran (2,3,7,8-TCDF) is one of the most toxic furans, and in this chapter we refer to it as "furan." Dioxins and furans evoke concern because they are typically toxic, they persist in the environment, and they bioaccumulate (that is, they build up in fat reservoirs and increase in concentration up the food chain, as in the much-publicized case of the pesticide DDT).

Concern about dioxin had arisen largely from its presence as a contaminant in herbicides, particularly in (2,4,5-trichlorophenoxy) acetic acid (or 2,4,5-T). Although the U.S. Department of Agriculture (USDA) canceled the registration of 2,4,5-T in 1970 for uses "around the home, recreation areas, and similar sites and in crops intended for human consumption," citing concern about possible adverse effects on developing fetuses, it was not completely banned until the EPA ordered an emergency suspension in February 1979 (EPA, 1981; EPA, 1987a). By 1983, more than 16,000 U.S. veterans who had served in Vietnam had filed disability claims for health problems (particularly cancer, rashes, and infertility) that they believed were related to their exposure to Agent Orange, which contained equal parts of dioxin-contaminated 2,4,5-T and 2,4-dichlorophenoxyacetic acid (or 2,4-D). Perhaps more frightening than the tragic stories of pain and suffering told by individual servicemen and their families, on *Nightline* Ted Koppel stated that "during all the years that dioxin has been sprayed in this country on forests, roadsides, and gardens, it's estimated that as many as 165 million Americans have been exposed to the chemical" (ABC, 1983).

But 2,4,5-T had emerged as only part of the dioxin problem. The *Nightline* program came on the heels of the federal government's decision to purchase an entire town in order to minimize human exposures to dioxin. The Centers for Disease Control (CDC) had blamed dioxin (in waste oil that was sprayed for dust control in Times Beach, Missouri) for the illnesses of two children and the deaths of more than sixty horses, seventy chickens, several dogs, a dozen cats, and hundreds of wild birds. Dioxin was also found at a number of abandoned hazardous waste sites, including the notorious and highly publicized Love Canal site in upstate New York. In addition, U.S. and Canadian studies had found dioxin in fish caught in the Great Lakes, leading several of the eight Great Lakes states (Illinois, Indiana, Michigan, Minnesota, New York, Ohio, Pennsylvania, and Wisconsin) to request health guidance from the Food and Drug Administration (FDA). Although it had no enforcement authority, the FDA responded by developing a "level of concern" guideline of 25 parts per trillion (ppt) in fish for the region (that is, FDA "saw no public health problem below 25 ppt"; Miller, 1983). By 1983, dioxin had been identified in emissions from municipal incinerators and from polychlorinated biphenyl (PCB) transformer fires, and it was also recognized as a by-product contaminant in some industrial chemicals, such as chlorophenols (EPA, 1987a).

Amid all of these concerns, in 1983 the U.S. Congress requested that the EPA conduct a national evaluation of the extent of dioxin contamination in air, water, and food. The EPA responded with the National Dioxin Study (EPA, 1983). In the process of measuring background concentrations of dioxin at sampling sites that were *not* thought to be directly affected by known sources of dioxins, the EPA made an unexpected finding: fish downstream from pulp and paper mills had some of the highest dioxin concentrations collected during sampling (EPA, 1987a). This finding was disturbing not only to the paper industry but also to the state and federal agencies responsible for protecting the public health. Following this discovery, the EPA and the paper industry collaborated in chemical sampling efforts that measured dioxin and furan in the pulp used to make a variety of consumer products and in wastewater and sludge from the mills.

Since 1985, the EPA, the FDA, the Consumer Product Safety Commission (CPSC), and various state agencies have performed numerous studies to estimate potential risks to human health and ecosystems from exposures to dioxin and furan released by pulp and paper mills. In 1990, for example, a comprehensive dioxin risk assessment of the 104 U.S. mills that produce bleached pulp evaluated approximately 120 different pathways of human exposure to dioxin and furan in mill sludges, effluents, and finished products (EPA, 1990a). The industry has been active in the risk assessment process, producing numerous assessments itself, as well as supporting research to resolve scientific uncertainties.

Risk assessments identified serious problems that the pulp and paper industry needed to address. The paper industry responded to the dioxin problem in part by contesting the EPA's toxicity evaluation of dioxin and its proposed regulations and permit limits, but also by investing more than a billion dollars to institute process changes that have dramatically reduced both the formation and the discharge of dioxin and furan. Since 1988 the industry has reduced the mass of dioxin and furan in its aggregate effluent emissions by more than 90 percent (NCASI, 1995).

Several factors that were influenced by qualitative and quantitative risk assessments (including publicity associated with news reports and fish advisories, consumer demand for dioxin-free products, and regulatory standards) contributed to the decisions of mills to make process changes (see Figure 7.1). This chapter explores the process of change in the paper industry and the role that risk assessment has played. We begin by describing the industry and the process used to make paper, including a discussion about the formation of dioxins and furans as unwanted by-products. Next, we discuss the health evidence that formed the basis for risk assessments, along with important aspects of the risk assessments for dioxin. Finally, we consider the importance of risk assessment in the regulations and industrial change, and the improvement in environmental quality that has resulted from these changes.

| Making Paper

Papermaking is one of the oldest industries, having been invented by the Chinese in 105 A.D. and standardized by the time the colonists arrived in America.[1] Making paper involves three basic steps: 1) making a solution of fibers suspended in water (called pulp) from wood, cotton, straw, waste paper, other vegetation, or recycled materials, such that the paper will have the desired properties (for example, brightness, color, and thickness), 2) matting the pulp on a fine-mesh screen that allows the water to escape and hydrogen bonds to form between the fibers, and 3) pressing and drying the mat after removal from the screen. Originally, old cotton and linen rags served as the main fibers for pulp, but with the invention of the printing press in the fifteenth century, paper demand quickly surpassed supply, and papermakers sought other sources of fiber.

The discovery of cellulose fibers in wood is traced back to the eighteenth-century French physicist René-Antoine Réaumur, who observed that wasps built paperlike nests. Following this observation, straw and different types of wood were used to make pulp, and pulpwood was first used commercially in 1844. Current paper production uses four different sources of fiber: hardwood (oak, maple, birch,

[1] This section relies on Weeks (1916), Luken (1990), Ohanian (1993), OTA (1989), and Rapson (1963), and specific references to these works are omitted.

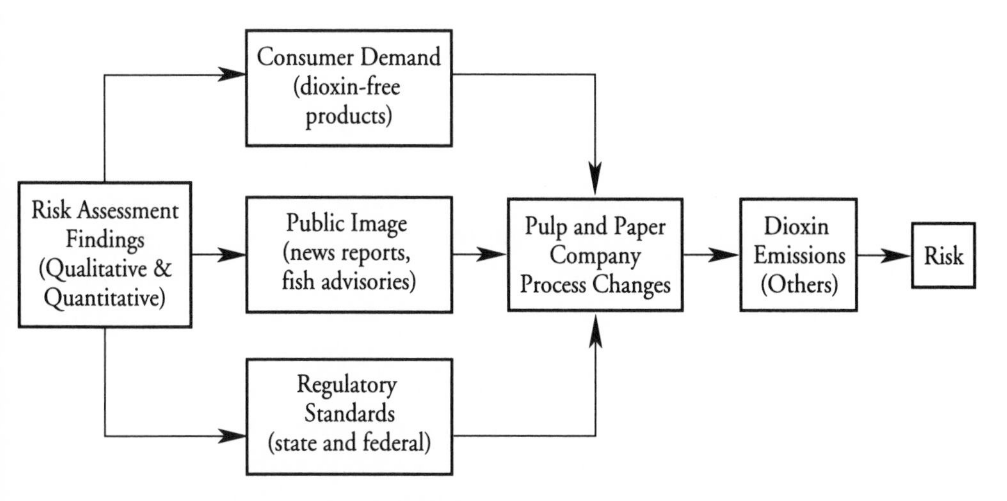

FIGURE 7.1 | Schematic for the influence of risk assessment on the pulp and paper industry.

beech), softwood (pine, spruce, hemlock), recycled paper or paperboard, and nonwood fibers (cotton, flax, hemp). With their longer fibers, softwood pulps are stronger and less dense than hardwood pulps, and about twice as much softwood pulp is produced in the United States.

The main challenge in using wood as a source of pulp is that it contains a light-absorbing material called lignin, hemicellulose, and extractives (turpentine, resins, tall oil, soap) in addition to cellulose fibers. Lignin, which holds cellulose fibers together in the raw wood, may discolor and weaken the final paper product if it is not removed. Unfortunately, removing lignin is difficult, and high rates of removal are accompanied by lower net pulp yields. Thus the goal of preparing pulp from wood for any particular product is to maximize the pulp yield, subject to constraints on the maximum amount of residual lignin allowable in the final pulp. Since hemicellulose and the other extractives are easier to remove, they are less of a concern, although typically they must be removed as well.

Techniques for chemical pulping were developed in the mid-1800s. The objective in chemical pulping is to dissolve a large fraction of the lignin using nonoxidizing chemicals (alkalis, sulfides, or sulfites) that are cheap, effective at high temperatures, and often recoverable for reuse. Different strengths and purities of pulp could be produced, depending on length of "cooking" time and the strength of the chemicals used for cooking. Commercially viable in 1866, the soda process involved cooking wood chips in a harsh solution of soda ash and lime to produce relatively weak pulp at a yield of approximately 40 percent. The sulfite process, viable in 1882, used a solution of sulfur dioxide and calcium bisulfite and produced a relatively light-colored pulp. The sulfate (or kraft) process, viable in the United States in 1909, used sodium hydroxide and sodium sulfide to create a very strong brownish pulp used to make bags, wrapping paper, container boards, and towels. In the United States, chemical pulping currently accounts for approximately 60 percent of pulp production, and approximately 95 percent of this is produced using the kraft process.

Simple mechanical pulping (grinding the wood) does not remove the lignin and other components, but it provides pulp yields of over 90 percent, and it currently accounts for approximately 7 percent of pulp production. The pulp produced is used for short-lived products like newspapers, tissue, catalogs, and throwaway molded items. Semi-chemical pulping, which combines chemical and mechanical treatments, accounts for 5 percent of pulp production. Recycled fibers, which account for approximately 28 percent of pulp production, are mainly processed mechanically, although they undergo several cleaning steps and possibly deinking processes as well.

| Brightening Pulp

Although pulping removes approximately 70 to 95 percent of the lignin (depending on the wood stock and the pulping process), longer and more severe pulping can significantly degrade the cellulose fibers. Residual lignin makes the pulp brownish in color, and consequently it is sometimes called "brownstock" in the industry. Approximately 55 percent of the chemical pulp produced is bleached to further remove lignin (a process called delignification), to decolorize any remaining lignin and to obtain pulp suitable for making paper and paper products of high brightness, longevity, and strength (for example, writing paper, paper towels).

Mills typically bleach pulp in a number of different stages that alternate between the use of bleaching agents (elemental chlorine, chlorine dioxide, sodium hypochlorite, hydrogen peroxide, oxygen) and a strong alkali (usually sodium hydroxide) to remove dissolved lignin from the fiber surfaces. The choice of bleaching agents is influenced by cost, selectivity for degrading lignin, and the associated degree of damage to cellulose. Different mills use different bleaching processes to meet a variety of product needs, but each process is designed to achieve the maximum bleaching efficiency while minimizing damage to or loss of cellulose. The brightness of the pulp is graded on a scale from 1 to 100, and several techniques exist for measuring brightness. Fully bleached pulp is used for white paper products and typically has a brightness of 85 to 90.

The powerful bleaching action of chlorine was recognized almost immediately after its discovery as an element in 1774, and chlorine-based bleaching agents have been used by the industry since 1804. Invented in 1799, "bleaching powder" (calcium hypochlorite) became the first chemical agent used to bleach pulp. Between 1900 and 1930, multistage bleaching techniques were developed. With the advent of large-scale production of liquid chlorine to meet the demands of the military in World War I, lower-cost elemental chlorine replaced hypochlorite in the first bleaching stage. Since unbleached pulp contains relatively high levels of lignin, relatively unselective bleaching agents (elemental chlorine, oxygen) can be used very effectively in the first bleaching stage. In addition to its relatively low cost, chlorination was preferred to hypochlorite for the first bleaching stage because of its rapid reaction with lignin at low temperatures, formation of lignin products that are soluble in dilute alkali solution, and minimal damage to cellulose. Hypochlorite continued to be used in later bleaching stages, because of its ability to brighten pulps and its limited degradation of cellulose. However, chlorine dioxide, which is highly selective for lignin, increasingly replaced hypochlorite as a bleaching agent. By the 1950s, pulp mills had learned how to generate and use chlorine dioxide on site; and by the early 1960s the majority of mills were using it in one or more bleaching stages. Due to its relatively high cost, chlorine dioxide did not readily replace the use of elemental chlorine as the first-stage bleaching agent.

Oxygen bleaching techniques were introduced in the 1950s and 1960s and were first commercially implemented in a South African mill in 1970. Oxygen delignification technology uses oxygen instead of chlorine in the first bleaching stage. It was embraced by mills in Sweden because it reduced their production and release of chlorinated organic byproducts into the widely contaminated Baltic Sea. In contrast, U.S. mills did not consider the relatively expensive oxygen delignification technology worthwhile, because they had adopted secondary biological waste treatment technology to reduce their releases of conventional pollutants like biological oxygen demand (BOD) into lakes and streams. Secondary treatment reduces concentrations of organic compounds released to air and water, but it may lead to the generation of a sludge waste stream that typically contains residual organic compounds.

In addition to oxygen delignification, efforts to use oxygen for bleaching have also focused on oxygen enrichment during alkaline extraction, use of ozone as a replacement for chlorinated agents, and elimination of hypochlorite by replacement with oxygen or peroxide. Pilot studies indicated that ozone may be an effective bleaching agent (Liebergott, 1983) and approximately fifteen kraft mills worldwide now use ozone in their bleaching sequence. While it is possible to produce pulp of high brightness and strength using ozone bleaching and a totally chlorine-free (TCF) process, this has not been demonstrated commercially to date. The development of oxygen-bleaching technologies has allowed some mills to produce bleached pulp of medium brightness using a TCF process, although the EPA recently concluded that "the limited range of papergrade TCF products currently produced and sold in the U.S. market indicates that TCF technology is not yet available to make the full range of products produced by [elemental chlorine free] or similar processes," (EPA, 1996:36839).

Factors that have played a role in locating mills have historically included proximity to water (to power the mill and to make the pulp solution), proximity to consumers (to minimize transportation costs), and availability of capital, skilled labor, and raw materials (for example, rags, waste paper, and pulpwood). Given these considerations, paper mills were historically located on waterways on the outskirts of population centers. Currently the industry is the largest U.S. industrial water user, and it produces on the order of 100 million tons of pulp annually. Since each American consumes an average of approximately 700 pounds of paper products per year, the U.S. is a net importer of paper, even though the U.S. is the largest worldwide producer of paper.

| Formation of Unwanted Chlorinated By-products

Using some bleaching agents can lead to the production of a wide array of chlorinated by-products, including dioxins, furans, chlorophenols, chloroform, and many other identified and unidentified compounds, most of unknown toxicity. The relative amounts and types of organochlorides produced depends on the nature of the bleaching agent and its reactions with brownstock. In particular, the use of elemental chlorine leads to substantially greater production of chlorinated by-products than the use of chlorine dioxide, while use of hypochlorite tends to lead to greater production of chloroform.

Concern about the wide array of chlorinated by-products emitted by mills primarily arose in Sweden, where studies conducted between 1977 and 1985 associated biological effects on aquatic life in the Baltic Sea with pulp and paper mills (Sodergren et al., 1988). These studies identified chlorinated organic compounds as the probable source of the impacts on aquatic life and further indicated that the hazard was lower for mills using oxygen delignification or increased substitution of chlorine dioxide for elemental chlorine (Environment Ontario, 1988). The Swedish Environmental Protection Agency (SEPA) was concerned that only a relatively small portion of the numerous chlorinated organic compounds were individually identified and that many of the unknown compounds were likely to be persistent in the environment and to cause adverse biological effects.

The U.S. data that associated dioxin with bleaching pulp added a new dimension to existing concerns about the overall formation of chlorinated by-products. During the National Dioxin Study, the EPA found dioxin in fish collected from 57 percent of the sites sampled downstream from pulp and paper mills. Mills that bleached wood pulp were subsequently implicated as possible sources of dioxin following the sampling of wastewater treatment sludge from mills in late 1985. Industry research efforts around the world then focused on determining the source of the dioxins and furans, finding ways to eliminate or treat them, and improving the scientific understanding of health effects caused by exposure.

The U.S. industry initiated research through the National Council of the Paper Industry for Air and Stream Improvement (NCASI) to verify the EPA's preliminary findings and to validate and improve the analytical procedures then in use. In 1986 the EPA began designing a sampling plan to study one bleached kraft pulp and paper mill in detail, to determine the source of dioxin. The industry, coordinated by the American Paper Institute (API), urged the EPA to expand the study to five mills and offered to cooperate in and partially fund the study. Conducted during 1986 and 1987, the Five-Mill Study (NCASI, 1988a) confirmed that chlorine-based bleaching led to the formation of dioxins and furans, which were detected in some samples of the effluents, sludge, and bleached pulp at the five mills. The study further demonstrated that dioxin and furan were the principal PCDDs and PCDFs found, and suggested that once formed they were not destroyed by downstream processing or wastewater treatment.

Based on the results of the Five-Mill Study, the EPA initiated actions to characterize the formation of dioxins and furans from all of the 104 U.S. mills that practiced chlorine bleaching. In late 1987, a cooperative agreement between industry and the EPA was reached to collect samples and process information from all 104 mills, with industry paying for the approximately $3 million study. The 104-Mill Study (NCASI, 1990a), conducted in mid to late 1988, provided nationwide estimates of the release of dioxin and furan from the industry into water, sludge, and bleached pulp, and characterized the relatively high amount of variability within the industry. Also in late 1988, the industry designed and conducted an intensive study of twenty-two bleaching lines to relate the formation and distribution of dioxin and furan to such bleach plant operating conditions as levels of precursors, delignification practices, bleaching sequences, wood type used, mill location, and chemical use (NCASI, 1990b).

While these studies were being conducted, other industry research was identifying some factors that lead to the production of dioxin and furan, as well as some techniques to reduce their production. In 1988 Swedish and Canadian investigators (Kringstad and McKague, 1988; Axegard, 1988) recognized the importance of the ratio of applied chlorine in the first bleaching stage to the amount of lignin entering the stage in the production of dioxin and furan. The industry's intensive study of 22 bleach lines showed that production of dioxin and furan occurred primarily in the first bleaching stage and determined that it was correlated with the amount of elemental chlorine used (NCASI, 1990b). It also provided more support for the hypothesis advanced by Canadian researchers (Allen et al., 1988), that oil-based additives were precursors to the formation of dioxin and furan during bleaching (NCASI, 1990b).

The intensive study found, additionally, that mills using different technologies aimed at reducing the formation of dioxin and furan were having varying levels of success. The study included numerous technologies: oxygen delignification (Reeve, 1984; Kleppe and Storebraten, 1985); split addition of chlorine (adding small amounts of chlorine sequentially instead of the full charge at once) with simultaneous control of the mixture acidity (Hise, 1989); and high chlorine dioxide substitution for the first chlorine bleaching stage (using mostly chlorine dioxide and low levels of elemental chlorine) (Pryke and Reeve, 1985). Industry research subsequently suggested that a chlorine dioxide substitution level of at least 40 percent (for chlorine in the first stage) was required to reduce production of dioxins and furans below detection limits (Berry et al., 1989). In a separate study, the industry also investigated end-of-pipe treatment technologies for reducing dioxins and furans and found them to be relatively ineffective (NCASI, 1992).

A major challenge in attempting to understand and monitor releases of dioxin and furan from the industry has been developing techniques capable of reliably measuring them at very low levels. In the mid-1980s, fewer than five U.S. laboratories were capable of quantifying very low dioxin concentrations. The nominal minimum levels of detection for the EPA's proposed analytical method (method number 1613A) for measuring dioxin are 10 parts per quadrillion (ppq) in mill effluent and 1 part per trillion in sludges or pulps (NCASI, 1994). (10 ppq is 0.000000000000010 parts of dioxin per part of water, and it is approximately equivalent to 1 ounce in 1,562,000,000,000 gallons.) Although some mills may report detection limits as low as the method detection limits for clean water (1 to 3 ppq), and it may be possible to detect even lower levels with extensive sample preparation, the EPA considers 10 ppq to be the practical level of detection.

| Toxicity of Dioxin and Related Compounds

While the industry and scientists grappled with the need to detect and develop technology to eliminate low concentrations of dioxin and furan, biologists and toxicologists grappled with issues of understanding and characterizing the biological activity of dioxin.[2] The industry had little involvement in supporting or contesting most of the research because it was conducted prior to the discovery of dioxin releases from mills, but following the discovery, the pulp and paper industries were active in supporting development of both the science and policy.

Noncancer Effects

Consistent with its reputation as "the most toxic chemical known to mankind," dioxin's toxicity has been extensively studied and reviewed. The studies suggest that dioxins are capable of producing a wide spectrum of adverse health outcomes in animals, including chloracne, reproductive and developmental effects, immunotoxicity, cancer, and death. In humans, high doses of dioxin are related to chloracne, and there have been reports of alterations in testosterone levels, liver function, and glucose metabolism. Dioxin is ubiquitous in the environment, and consequently everyone is exposed to some degree. Environmental dioxin is believed to result primarily from human activities of the twentieth century, such as the use of chlorinated compounds (for example, herbicides and PCBs) and the production of incinerator and diesel emissions (SAB, 1995). Higher exposure levels are associated with certain jobs (such as work in some chemical production facilities), industrial accidents (for example, the accidental release of approximately one pound of dioxin following a reactor failure at a trichlorophenol plant in Seveso, Italy, in 1976), some hazardous waste sites (such as the Love Canal and Times Beach Superfund sites), and emissions from facilities that generate dioxins as contaminants in their production processes.

Animal studies have shown that dioxin is highly toxic to every species tested, although there are large species differences in the degree of sensitivity. The acute lethality of dioxin, as measured by the single dose that is lethal to 50 percent of the studied population (LD_{50}), ranges from 0.6 μg/kg (or 0.0000006 grams/kg body weight) for male

[2] This section relies on EPA (1985), EPA (1988), and EPA (1994), and specific references to these works are omitted.

guinea pigs to more than 5,000 μg/kg for male hamsters and Han/wistar rats, although most of the measurements are on the order of 100 μg/kg (EPA, 1985; Pohjanvirta, Unkila, and Tuomisto, 1993). Lethal poisoning is typically characterized by thymic atrophy and severe weight loss (which may cause the animal to "waste away"), and death may occur up to forty days following exposure. Exposure to dioxin produces acute liver injury in rats, mice, and rabbits. Immunosuppression has also been observed in mice. Subchronic or chronic exposure in rats and mice severely affects the liver and may lead to systemic hemorrhage, edema, and suppressed thymic activity.

Although early research showed that dioxin was a specific cause of chloracne in humans at high levels of exposure (Kimmig and Schulz, 1957), insufficient data are available to determine the doses at which chloracne will occur. In 1969 and 1970, studies of developmental toxicity demonstrated that 2,4,5-T contaminated with dioxin led to adverse effects on fetal development in rats and mice, including increased fetal deaths (Courtney et al., 1970). Dioxin exposure has produced cleft palates and hydronephrosis in mice, a higher incidence of extra ribs in rabbits, and edema, hemorrhage, and kidney anomalies in rats. The most sensitive developmental effects are those on male and female reproduction systems in rats and object learning behavior in monkeys (Peterson, Theobald, and Kimmel, 1993). Exposed male rats show decreased testicular and accessory sex organ weights, abnormal testicular morphology, decreased spermatogenesis, and reduced fertility. Effects of exposure on female rats include reduced litter sizes and fertility, and changes in the female gonads and menstrual cycle. In 1989 the EPA characterized dioxin as "the most potent reproductive toxin yet evaluated by EPA," based on results in laboratory animals (EPA, 1989a), and the suggestion that dioxin might mimic or interfere with natural hormones (such as estrogen) has made it to the popular press in the book *Our Stolen Future* (Colburn, Dumanoski, and Myers, 1996).

Concern about immunotoxicity from exposure to dioxin stems from the recognized importance of the immune system in resisting and controlling disease. Animal studies have demonstrated that exposure to very low levels of dioxin may alter immunological function, although no particular syndrome has emerged from the literature. Dioxin and related compounds have been shown to alter the effects of several endocrine hormones that regulate immune response (including glucocorticoids, sex steroids, thyroxine, growth hormone, and prolactin). Despite considerable effort by scientists, the mechanism by which dioxin causes immunosuppression in animals remains unknown.

Cancer

In spite of the myriad effects produced by dioxin and related compounds, cancer has been the health effect of greatest regulatory concern. Although several animal cancer bioassays have been conducted (Van Miller, Lalich, and Allen, 1977; NTP, 1982, Rao et al., 1988), a study conducted by Dow Chemical Company researchers has had the largest regulatory impact (Kociba et al., 1978). In the Kociba study, rats were exposed orally to dioxin at levels of 0.1, 0.01, and 0.001 μg/kg/day over a period of two years, and the survival rates of the rats in the upper two dose groups were relatively poor. The rats treated with dioxin exhibited a significantly higher number of hepatocellular carcinomas (malignant liver tumors, females only), stratified squamous cell carcinoma of palate or nasal turbine (males and females), keratinizing squamous carcinoma of the lung (females only), stratified squamous carcinoma of the tongue (males only), and adrenal cortical adenoma (males only) than the control group (EPA, 1985). At the same time, the treated animals exhibited a lower number of mammary tumors, suggesting that dioxin might also have protective effects.

Epidemiological evidence for dioxin is extensive and suggestive, but inconclusive. The main studies can be grouped into occupational studies, studies of community residents exposed inadvertently, and studies of Vietnam War veterans (EPA, 1994). Human exposure to dioxin may produce a significantly elevated risk of soft tissue sarcomas (a relatively rare tumor), respiratory cancer, stomach cancer, and non-Hodgkins lymphoma, although some studies show elevated risks for these outcomes while others do not, and the evidence is disputed. Exposure could also have a protective effect, since the number of observed cases of breast cancer and endometrial cancer were lower than expected in one cohort (Bertazzi et al., 1993). However, interpretation of the existing epidemiological evidence is complicated by confounding factors. Overall, the evidence has not proven adequate to estimate the health risk of dioxin exposure for humans, and scientists have turned to mechanistic studies to attempt to resolve outstanding scientific uncertainties.

Mechanisms of Action

In humans, dioxin is readily absorbed through the gastrointestinal tract, and it is metabolized in the liver via the cytochrome P-450 system of enzymes. However, since dioxin is a relatively poor substrate for metabolism and it collects in fat, it tends to persist in the body, and it has a half-life on the order of ten years (Pirkle et al., 1989).

In the 1970s, Alan Poland of the University of Wisconsin initiated the first studies on mechanisms for dioxin's toxicity. Its unusually high potency suggested the possibility of the existence of a receptor for dioxin. Using radiolabeled dioxin as a ligand, Poland and colleagues found that the dioxin bound saturably, reversibly, and with high affinity to an intracellular protein (Poland, Glover, and Kende, 1976). They called this protein the aromatic hydrocarbon (Ah) receptor because of its ability to bind and mediate the response to other aryl-hydrocarbon molecules. The Ah receptor was further implicated in a broad spectrum of biochemical, morphological, immunological, neoplastic, and reproductive effects that dioxins and furans elicit. In other studies scientists identified one important feature of the ligand-bound Ah receptor: that it does not bind strongly to DNA by itself. Thus, the receptor must interact with at least one other factor to elicit a response.

In the early 1980s, scientists identified a second protein, called the Ah receptor nuclear translocator (Arnt), named for the role that the protein plays in translocating the liganded receptor complex from the cytoplasm into the nucleus and in mediating the binding of the complex with DNA and the activation of gene transcription. Studies have confirmed that the Arnt protein does not bind to dioxin or DNA in the absence of the Ah receptor protein. Thus, adverse effects of dioxin exposure appear to result from alterations in gene expression in the presence of these two proteins.

Other proteins, some yet to be identified, may play a role in the gene regulatory effects of dioxin and in the activation of gene transcription by dioxin. Complicated responses (like cancer) are likely to involve multiple genes, events, and environmental factors. However, in 1990 scientists recognized one apparent similarity for all effects: dioxin must bind to and activate the Ah receptor (Banbury, 1991). The exact number of receptors that must be bound in order to produce an effect is unknown, but it is likely that several hundreds or thousands of receptor sites must be occupied. Consequently, this type of mechanism suggests that there may be a dose or threshold below which adverse effects will not occur. An example of this is suggested by the ability of the dioxin-Ah-Arnt complex to increase activity of the cytochrome P-450 enzyme system following its transport into the nucleus. In this case, no toxic effects are known to occur at levels below the level required for enzyme induction (Roberts, 1991). Nonetheless, the possibility exists that dioxin may produce toxic responses that are not mediated through the Ah receptor (SAB, 1995) and perhaps more important, the threshold is irrelevant if body burdens associated with background exposures already exceed the threshold.

Mechanistic studies have provided insight into observations about the inconclusive mutagenicity studies and have demonstrated dioxin's ability to promote the carcinogenic process, even if it does not initiate tumors. Since dioxin does not bind appreciably to DNA directly, it is unlikely to be a mutagen, although it does act as a tumor promoter. Studies have confirmed the presence of the Ah receptor in all animals studied, and several functional forms of the Ah receptors have been identified in many human tissues (including liver, lung, lymphocytes, and placenta). Nonetheless, very little is known about the linkage between the Ah receptor and specific cellular toxicity. Several key issues must be addressed in characterizing risk using a "threshold" model for dioxin: some humans may be more susceptible than others to dioxin and consequently may have different response thresholds; background levels of dioxin must be characterized, since the baseline exposures may be very close to or even above the threshold; and levels of other compounds (including natural ligands) that react in the same pathways must also be considered. Research on these issues is ongoing.

Accounting for All of the Dioxins and Furans

Since dioxins and furans tend to be coproduced and they are believed to elicit the same biological responses, efforts are made to account for the total effective exposure using a "toxic equivalency scale." Results of laboratory studies can be used to compare the toxicity of various dioxins and furans to dioxin (the most potent of the group). For each of the other dioxins that are thought to be carcinogenic based on their structures and on in vivo and in vitro tests, toxicity equivalence factors (TEFs) are used to represent the toxicity of each compound relative to dioxin, which is assigned a TEF of 1 (e.g., furan has a TEF of 0.1). The scale (albeit dated) of TEFs used by the EPA reflects international consensus and is based on considerations about binding as well as effects other than cancer (EPA, 1989a). Dioxin toxic equivalents (TEQs) are found by multiplying the appropriate TEF by the concentration of each dioxin-related compound, and then adding the sum of these to the amount of dioxin. The use of TEQs allows a larger number of potentially active compounds to be included in a risk assessment, but it also adds substantial uncertainty into the process, since the TEFs are based on relatively little data. For example, only one of the other seventy-five dioxins has been tested for carcinogenicity in an animal bioassay.

| Predicting Human Cancer Risks

The EPA's attempts to analyze the cancer dose-response relationship for dioxin have been characterized by multiple trips to the agency's Science Advisory Board, much as was the case with perchloroethylene, or perc, described in Chapter 4. Unlike the perc case, however, the classification of dioxin as a carcinogen was a relatively minor controversy, while derivations of quantitative dose-response relationships was a major controversy.

In the early 1970s concerns about dioxin were based on adverse developmental effects, but in the late 1970s regulatory concern shifted to cancer, following the release of cancer bioassays. In 1981, the EPA's Carcinogen Assessment Group (CAG) first classified dioxin when it concluded that the human studies provided "strongly suggestive" evidence for the carcinogenicity of 2,4,5-T or dioxin, or both, and that the animal studies provided "substantial" evidence that dioxin is carcinogenic to mice and rats (EPA, 1981: 88, 100). The CAG also indicated that, based on the liver tumor results of the study by Kociba and colleagues (1978), dioxin appears to be one of the most potent carcinogens known.

Since mechanistic data on dioxin were relatively limited at that time, the EPA chose to rely on its default dose-response model, which is linear at low doses (that is, any dose above zero is considered to pose a finite and additive risk). The CAG assumed that dioxin could be a complete carcinogen based on two arguments: 1) that metabolism of dioxin could produce an intermediate that would react with DNA, and 2) that dioxin's chemical structure made it "likely that it could intercalate into DNA and also act as a genotoxic carcinogen" (EPA, 1981: 89).

Prior to its dose-response modeling efforts, the CAG had the histopathological slides from the Kociba and colleagues' (1978) study reevaluated by Robert Squire, a pathologist at Johns Hopkins University Medical School. Using its standard procedures, the EPA's CAG derived an oral cancer potency factor of 4.25×10^5 $(mg/kg/day)^{-1}$ for dioxin based on Squire's interpretation of the slides for female rats. This potency factor was approximately two times higher than the one derived using the original study's diagnoses (Kociba et al., 1978).

In 1984 and 1985, the EPA revised its dose-response assessment as described in its *Ambient Water Quality Criteria* (AWQC) report (EPA, 1984) and in a health assessment document (HAD) (EPA, 1985). In the HAD, the EPA indicated that the available data for mutagenicity were "inconclusive"; animal evidence was judged "sufficient," since dioxin produced a statistically significant increase in number of tumors for two species of animals, and human evidence was "inadequate" (EPA, 1985: 2–7, 2–8). According to the EPA's weight-of-evidence classification criteria, dioxin was placed in category B2, a "probable" human carcinogen (EPA, 1985: 2–8). The EPA determined that the available mechanistic and pharmacokinetic data were "insufficient" to support deviation from the EPA's default dose-response model (EPA, 1985: 11–136). The cancer potency factor did change, however, from the 1981 estimate. In the HAD and the AWQC, the potency estimate was changed to 1.56×10^5 $(mg/kg/day)^{-1}$ because it included an adjustment for the early mortality observed in the Kociba et al. (1978) study, and it used the geometric mean of the results obtained based on the original pathology and Squire's review.

In the HAD, the EPA also substantiated its claim that dioxin was the most toxic compound evaluated. By comparing potencies for all of the fifty-five chemicals that the EPA had evaluated to date, dioxin was shown to have a potency that was approximately two orders of magnitude higher than the next most toxic compound, and orders of magnitude higher than the known human carcinogens benzene and vinyl chloride (EPA, 1985).

Perspectives of Different Agencies, States, and Countries

The EPA's position on dioxin's cancer potency differed from those of other agencies. In its development of a level of concern for dioxin in fish in the Great Lakes, the FDA used a potency factor of 1.75×10^4

$(mg/kg/day)^{-1}$. This number, which is almost an order of magnitude smaller than the EPA's factor, differed from the EPA's in four ways: 1) the FDA extrapolated from rat to man on the basis of body weight (the EPA used surface area); 2) the FDA used the original pathology (the EPA also uses Squire's results); 3) the FDA did not adjust for early mortality; and 4) the FDA assumed an average human body weight of 80 kg (the EPA uses 70 kg). The CDC derived a slightly higher cancer potency factor of 3.6×10^4 $(mg/kg/day)^{-1}$ in response to the dioxin contamination at the Times Beach, Missouri, Superfund site. This number differed from the EPA's in four ways: 1) the CDC extrapolated from rat to man on the basis of body weight; 2) the CDC used Squire's results only; 3) the CDC did not adjust for early mortality; and 4) the CDC used liver concentration at terminal sacrifice (instead of administered dose, which was used by the EPA). The three agencies met in January 1984 to discuss these differences but did not resolve them and instead agreed that the differences were "within the range of uncertainty inherent in the risk assessment process" (EPA, 1984: C-184).

By 1988 a wide range of dose-response assessments had emerged, and the EPA released a draft document that reexamined the scientific basis and methods used for its cancer dose-response approach for dioxin (EPA, 1988a). To facilitate comparisons between different assessments (some based on low-dose linearity and some based on a traditional toxicological reference dose [RfD] approach), the EPA defined the risk-specific dose (RSD) as the dose that yielded a particular upper-boundary estimate of incremental lifetime cancer risk (for example, 1×10^{-6}) when combined with the cancer potency factor. Since both the RSD and RfD could be interpreted as the lifetime daily dose likely to pose an insignificant risk, they can be considered on the same scale. For its analysis, the EPA specified a risk of 1×10^{-6}. Thus, dividing 1×10^{-6} by the cancer potency factor of 1.56×10^5 $(mg/kg/day)^{-1}$ and performing some unit conversions gives an RSD of 0.006 pg/kg/day.

Figure 7.2 shows the RSDs for three different risk levels (10^{-4}, 10^{-5}, 10^{-6}) and the RfDs for several agencies, states, and countries (based on EPA, 1988a). When 1×10^{-6} is specified as the risk level for the RSD, then the range of RfDs and RSDs spans three orders of magnitude, with the RSDs being much lower than the RfDs. (However, if 1×10^{-4} is specified as the risk level for the RSD, then the

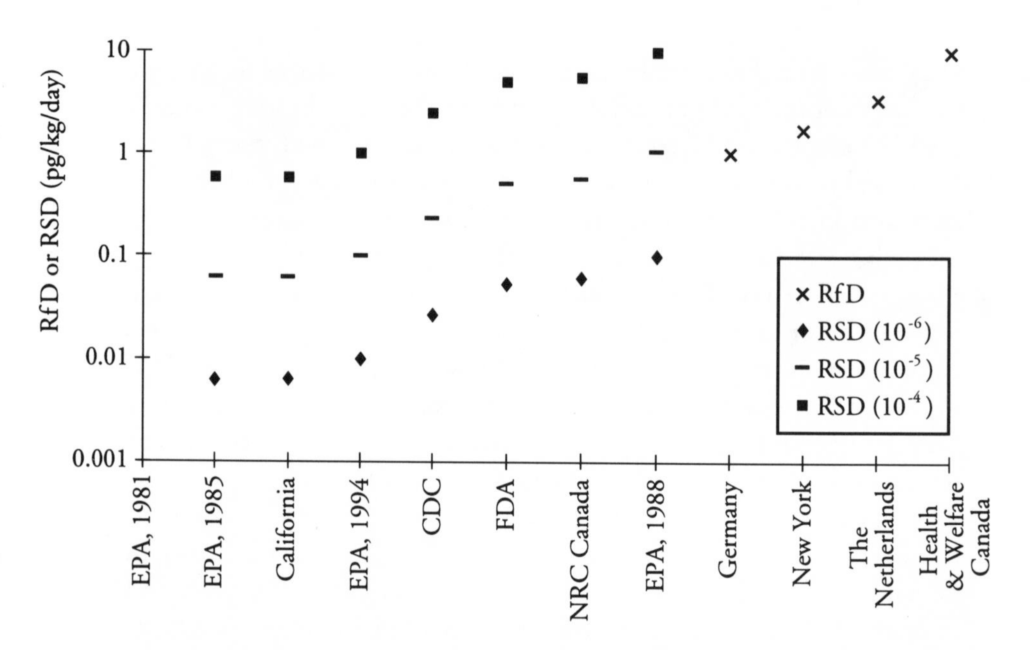

FIGURE 7.2 | Differences in dioxin potency between agencies, states, and countries. (RFD = reference dose; RSD = risk-specific dose for the risk levels given in parentheses.)
Source: EPA, 1988a.

RSDs and RfDs are in approximately the same range.) This wide range of values led to criticism by industry representatives, who claimed that the EPA was taking an extreme position.

Although more mechanistic data about dioxin had been generated since 1985, the 1988 EPA draft found that the "enormously rich data base" on dioxin did not provide a sufficient basis for deviation from the default low-dose linear model. However, the draft indicated that while the work group charged with preparing the draft was not convinced that the default model was appropriate for dioxin, it had used the default because there was no better alternative. The work group also used several "science policy" considerations to adjust the cancer potency factor downward to 1×10^4 (mg/kg/day)$^{-1}$, which increased the RSD to 0.1 pg/kg/day for a risk level of 1×10^{-6}:

- the scientific data indicate that the [EPA's] current upper bound for [dioxin] may be an overestimate;

- the scientific data do not permit an estimate of the extent of the overestimate;
- all of the . . . RSD estimates generated by the Federal agencies are arguably of equal scientific merit at this time;
- for strictly policy purposes, there is great benefit in Federal agencies adopting consistent positions in the absence of compelling scientific information; and
- an order of magnitude estimate of the RSD (potency), as opposed to some more precise estimate of the risk-specific dose, helps to convey the notion that the numerical expression is only a rough estimate (the science permits no greater accuracy). (EPA, 1988a: 51)

Finally, the work group recommended that the science and policy for dioxin should be reevaluated regularly in light of the extraordinary amount of research on dioxin and the likelihood that new relevant information would be released.

The EPA's Science Advisory Board (SAB) reviewed the 1988 draft and agreed with the work group that the linear dose-response model was inadequate and that alternatives were lacking. The SAB recommended that the EPA develop a mechanistic receptor-based model to generate a new potency estimate, and indicated that such an approach would not necessarily lead to a less stringent RSD (SAB, 1990).

In October 1990, an important shift occurred in the scientific debate about dioxin. At a conference held at the Banbury Center at New York's prestigious Cold Spring Harbor Laboratory, thirty-eight of the leading dioxin scientific experts from the United States, Canada, and Europe reached a number of general agreements:

- humans and experimental animals respond to dioxin similarly;
- effects in humans can be anticipated by effects in experimental animals, including enzyme induction, immunotoxicity, reproductive toxicity, developmental toxicity, and carcinogenicity;
- certain chemicals which are isostereomers of dioxin may behave the same as dioxin. These include certain polychlorinated and polybrominated dibenzofurans, polychlorinated and polybrominated dibenzo-p-dioxins, and co-planar chlorinated biphenyls; and

- all toxic effects of dioxin are mediated by the chemicals binding to a protein receptor within the cell cytoplasm; therefore a receptor-based risk assessment model is appropriate and should be developed. (EPA, 1991a: 2–3)

Consensus at the conference broke down over the implications of using a receptor-based model. Some scientists believed that dioxin would be far less risky with the new model, while others warned that such speculation was wrong or premature (Roberts, 1991). However, as one of the conference organizers indicated, the new approach involved much more than dioxin, because there was a sense that if the EPA could not develop a better alternative than its default model for a compound on which it had a large amount of information, then it probably could not do so for anything (Roberts, 1991).

In January 1991, shortly following this meeting, the National Institute for Occupational Safety and Health (NIOSH) released the results of a comprehensive cancer mortality study (Fingerhut et al., 1991). The study, published in the *New England Journal of Medicine,* reviewed all of the mortality records of U.S. chemical workers exposed to dioxin between 1942 and 1984. The final cohort included 5,172 men from twelve different facilities, and the dioxin exposures were relatively well characterized. The study divided the cohort into high- and low-exposure subgroups (expected to be exposed to levels approximately 500 and 90 times higher than the general population, respectively), largely based on duration of employment in a dioxin-contaminated job. The study found no increase in the risk of cancer for the low-exposure group and significantly increased numbers of soft tissue sarcomas and respiratory cancers in the high-exposure group. Although interpretation of the results is limited, due to the statistical power of the study and because of the inability to control for possible confounding by smoking and exposure to other chemicals, the EPA indicated that the study was significant because it was consistent with the status of dioxin as a carcinogen and "it [was] a study of human carcinogenicity [that lent] support to the use of a biologically-based dose-response model" (EPA, 1991a: 2).

At the time, the EPA found the Fingerhut et al. (1991) study, in conjunction with the Banbury conference and urging from the paper industry and other groups, compelling enough to initiate a scientific reassessment of dioxin (EPA, 1991a). The agency announced that its reassessment would include a biologically-based dose-response model for dioxin. As part of its effort to create a biologically-based model, researchers at the EPA and the National Institute of Environmental Health Science (NIEHS) began studies to characterize an appropriate threshold for humans. The results of these studies suggested that enzyme induction was likely to be occurring at existing background exposure levels and further indicated that a linear dose-response model might not be inconsistent with the observed results (Tritscher et al., 1992; Portier et al., 1993). This represented a turning point in the reassessment, because it offered strong support to the argument that a biologically based model may not result in lower estimates of risk and instead suggested that the potency estimate given in the 1988 draft HAD might not have been protective enough.

When the EPA released the new draft HAD in August 1994, the press picked up on this message with headlines like "New Report from EPA Strengthens Link Between Dioxin and Cancer, Other Ills," which appeared in the *Wall Street Journal* on September 13, 1994. After extensive reanalysis of the data, the EPA ultimately still relied on its default dose-response model at low doses, and it provided a cancer potency factor of 1×10^5 $(mg/kg/day)^{-1}$, which was ten times higher (leading to an RSD ten times lower) than the one proposed in the 1988 draft (see Figure 7.2).

In 1995 the SAB reviewed the draft HAD (SAB, 1995) and indicated that while the EPA had done an admirable job reviewing the scientific literature, its exclusive reliance on the default linear model and "relatively poor exposition of important points" weakened the assessment. The SAB recommended that the EPA consider alternative risk models that would be more consistent with the "body of available physiological . . . modeling of factors such as deposition, tissue dose, and excretion, as well as the epidemiological, and bioassay data" (SAB, 1995:3). The SAB also suggested that improved characterization of the noncancer risks would be required to facilitate meaningful analyses. The SAB indicated that EPA needed to do a better job presenting information about the continuum of effects that could occur at different exposures.

Overall, the task of characterizing the quantitative risks from dioxin has been contentious and difficult. The EPA did not attempt to change its dose-response model in the 1994 draft HAD as some scientists had expected. In spite of the large amount of toxicological information about dioxin, EPA has consistently relied on its default low-dose linear model for cancer as the basis of its risk assessments. Consequently, the "tremendous uproar from environmental groups and Congress" (Roberts, 1991) that some expected to accompany a change in the EPA's model has not occurred. However, given the SAB's comments, it appears that the EPA may face the task of additional model development for dioxin, and that addressing the controversy has mainly been postponed.

| Risks and Regulations for the Industry

In spite of the uncertainties about dioxin's cancer-causing potential, the EPA and other agencies have used quantitative estimates of cancer potency to establish risk-based criteria for dioxin and to perform quantitative risk assessments for each type of release from mills (effluent, sludge, and pulp). In numerous contests throughout the country, environmentalists and industrial advocates clashed over how stringent to make the regulations of dioxin.

Risk-Based Water Quality Criteria

The first national risk-based criteria for dioxin were published in the EPA's 1984 *Ambient Water Quality Criteria* report, which had no immediate impact on the pulp and paper industries because it was issued prior to the discovery of dioxin in mill effluent. Developed under Section 304(a)(1) of the Clean Water Act (CWA), these criteria were nonbinding guidelines for states on how strictly to regulate "toxic" pollutants under Section 307(a)(1). However, when adopted as state water quality standards under Section 303(c)(2) of the CWA, the state's criteria became enforceable maximum acceptable levels of dioxin in rivers and lakes.

In 1984, the EPA's policy was to establish separate water quality criteria for protection of aquatic life and human health when information allowed. In the case of dioxin, the agency found insufficient information to derive a national criterion for protection of aquatic life. Based on the ("nonthreshold") assumption that exposure to any amount of dioxin could increase cancer risk, the EPA indicated that *zero* represented the ambient water concentration that gave the maximum protection from potential carcinogenic effects due to dioxin exposure.

However, the EPA recognized that zero might not be attainable and consequently recommended that states consider risk-based criteria based on exposure scenarios involving ingestion of contaminated water and fish. Table 7.1 shows the numerical criteria that the EPA recommended for three risk levels. The lowest concentrations (and therefore the most stringent) were associated with combined ingestion of fish and water, although the risk came predominantly from ingestion of contaminated fish. In offering a range of numerical risk-based criteria (resulting from a range of acceptable risk levels), the EPA implied that states could exercise discretion as to the precise level of protection sought. Since all states chose a risk level of either 10^{-5} or 10^{-6}, the applicable guidance number for fish and water combined was either 0.13 ppq or 0.013 ppq. In its discussion of the numerical criteria, the EPA also noted that these levels were below the detection limit for dioxin and that states should consider the ability to detect dioxin when establishing enforceable maximum acceptable levels (EPA, 1984).

Efforts of Environmentalists to Stimulate Regulation

The Five-Mill Study conducted in 1987 armed environmentalists, who were concerned with the "soup" of chlorinated organic compounds discharged from mills, with a significant weapon (van Strum and Merrell, 1987). At that point, environmentalists began to pressure state regulators to revise the permits for paper mills to include a numerical limit for dioxin in effluent, as was already required by the Clean Water Act. Since the EPA had already issued the 1984 AWQC for dioxin — a guidance number — a numerical limit existed that could be applied to paper mills at the discretion of state regulators. In spite of this pressure, most states were reluctant to change the terms of permits prior to their

TABLE 7.1 | Ambient water quality criteria in parts per quadrillion (ppq) for different risk levels

RISK LEVEL	WATER AND FISH	FISH ONLY	WATER ONLY
1×10^{-5}	0.13	0.14	2.2
1×10^{-6}	0.013	0.014	0.22
1×10^{-7}	0.0013	0.0014	0.022

Note: Assumes cancer potency factor of 1.56×10^5 (mg/kg/day)$^{-1}$, consumption of 6.5 grams/day of fish and 2 liters/day of water, and a bioconcentration factor of 5,000 for fish.
Source: EPA, 1984.

expiration. In fact, states were also slow to adopt numerical criteria for dioxin (only New York state had adopted a standard for dioxin prior to 1987).

Nonetheless, the prospect that binding, risk-based dioxin limits would be added to permits became a reality for the industry following the 1987 amendments to the Clean Water Act. Due to the slow pace of action by states on all toxic pollutants (including dioxin) under the CWA, in 1987 Congress amended Section 303(c)(2)(B) of the Water Quality Act as follows: "Each state shall adopt criteria for all toxic pollutants listed pursuant to section 307(a)(1) of this Act for which criteria have been published under section 304(a), the discharge or presence of which in the affected waters could reasonably be expected to interfere with those designated uses adopted by the State, as necessary to support such designated uses. Such criteria shall be specific numerical criteria for such toxic pollutants."

Congress also created a timetable for adopting numerical criteria and thus created a clear mandate for states to control toxic substances (including dioxin) in effluents. In addition, the 1987 amendments to Section 304(1) set up a mechanism requiring states to identify problem areas in surface waters and to develop strategies to clean them up. Specifically, by February 1989 states were required 1) to submit a list of the specific industrial sources that contributed to "toxic hot spots" and the amount of each toxic substance discharged by each source, and 2) to submit an individual control strategy that would lead to the achievement of water quality standards within three years of adoption. The EPA moved on this mandate with fervor, issuing guidance in September 1987 calling for states to submit preliminary plans and lists (Houck, 1991). The EPA mandated that pulp and paper mills appear on the 1989 list of toxic hot spots and required the imposition of binding restrictions on dioxin discharges by 1992, as described below.

Environmentalists used legal tactics to pressure the EPA at the federal level. In 1985, before pulp and paper mills were recognized as a source of dioxin, the Environmental Defense Fund (EDF) and the National Wildlife Federation sued the EPA under the provisions of the Toxic Substances Control Act (TSCA) in an effort to move the EPA along in its assessment of the national dioxin problem and to stimulate regulatory action. Just prior to settlement of the suit, pulp and paper mills became a target, and representatives of the industry became interested parties in the 1988 consent decree that settled the suit out of court.

With respect to the pulp and paper industry, the consent decree includes two important provisions. First, the EPA agreed to issue guidance to permit writers regarding the effluent limits of mills by April 30, 1990. The guidance had to "address the permit issuance process and include a discussion of scheduling of and if feasible target dates for permit issuance" (Consent Decree, 1988:11). In addition, the EPA had to consider including a provision in the guidance that recommended monitoring fish downstream of mills with effluent levels below detection limits. Second, the EPA agreed to complete a "multiple pathway risk assessment considering sludges, water effluent, and products made from pulp . . . [considering] both occupational and nonoccupational risks, including but not limited to risks to aquatic organisms and from consumption thereof by wildlife and humans" by April 30, 1989 (Consent Decree, 1988:7). Following the risk assessment, the agency had the option to 1) refer the problem to another federal agency with appropriate regulatory authority, 2) decide that the risks were not unacceptable, 3) take up to a year to propose regulations, or 4) determine that more information was required prior to acting and propose a schedule to obtain the necessary information (Consent Decree, 1988).

This consent decree was very important, because it established time limits for the completion of agency guidance to states regarding water quality permits and indirectly on designation of toxic hot spots, performance of an integrated risk assessment, and initiation of agency action to address significant risks. Such time limits made it unlikely that the risk assessment would degenerate into a process of "paralysis by analysis."

Mills Identified as Sources of "Toxic Hot Spots"

One month after the consent decree was signed, the EPA issued an interim strategy for regulating pulp and paper mills that called for "aggressive action" on dioxin (EPA, 1988b). The strategy sought to eliminate dioxin discharges from the industry by requiring states to adopt and enforce water quality standards for dioxin. The EPA indicated that it would apply its own criteria, given in Table 7.1, for those states that failed to act. In the same document, which followed release of the 1988 draft health assessment document (EPA, 1988a), the EPA indicated that it was considering adoption of a new cancer potency factor for dioxin and thus opened the door for states to use different potency numbers. Overall, the EPA sent the message to states that they had a fair amount of discretion in adopting a dioxin standard, although standards still had to be approved by the EPA.

While the states were in the midst of proposing numerical standards for dioxin in water, in June 1989 the EPA finalized its guidance on Section 304(1) relating to "toxic hot spots" (EPA, 1989b). The CWA defines toxic hot spots as waters expected to remain polluted by toxic chemicals even after dischargers have installed the best available treatment technologies required by law. The EPA emphasized that individual control strategies would rely on effluent limitations that would be enforceable numbers included in water discharge permits. By mid-1989, all states except Arizona had submitted acceptable proposed lists of toxic hot spots and plans for action. The EPA compiled the national list and found that approximately 10 percent of the nation's waterways (17,365 of them) were not expected to meet the water quality standards. In addition, eighty-eight out of ninety-eight identified paper mills were listed, and dioxin limits were expected for them (Houck, 1991).

Since the numerical criteria for water quality standards were risk based, states could choose between the numbers (given in Table 7.1) by choosing a risk level, or adopt a different standard by relying on different technical assumptions. For each of the twenty-eight states that have bleached chemical pulp and paper mills, Table 7.2 lists the number of mills in the state, the numerical standard(s) adopted, the date of adoption, and the assumptions and risk levels used (EPA, 1993a). The adopted standards range over three orders of magnitude, as shown in Figure 7.3, which plots the strictest (lowest) standard for each state versus the date of adoption. Figure 7.3 shows a cluster of standards around 1 ppq adopted between 1990 and 1992 (following the release of the Section 304[1] lists), which are noticeably more permissive than the EPA's 1984 criteria. Ten of the eleven standards in this cluster correspond to southern states (the eleventh being New Hampshire).

In the South, Georgia started the trend by backing off its initially adopted October 25, 1989, standard of 0.013 ppq. On December 6,

TABLE 7.2 | Ambient water quality criteria for states with pulp and paper mills

STATES WITH MILLS	NO. OF MILLS	WATER + FISH (PPQ)	FISH ONLY (PPQ)	WATER ONLY (PPQ)	DATE ADOPTED	RISK LEVEL IF NOT 1×10^{-6}	RATE OF FISH CONSUMPTION IF NOT 6.5 GRAMS/DAY	BIOCONCENTRATION FACTOR IF NOT 5,000	CPF $(MG/KG/DAY)^{-1}$ IF NOT 156,000
NY[a]	2	—	1.0	—	9/1/85	na	na	10,000	na
OR	3	0.013	0.014	—	8/28/87	—	—	—	—
MT	1	0.013	0.014	—	9/1/88	—	—	—	—
ME	7	0.013	0.014	—	2/1/89	—	—	—	—
WI	8	0.03	—	—	3/1/89	1×10^{-5}	20.0	—	—
PA	5	0.01	—	—	3/25/89	—	—	—	—
NC	4	0.013	0.014	—	7/13/89	—	—	—	—
OH	1	0.13	0.14	—	2/1/90	1×10^{-5}	—	—	—
MD	1	—	1.2	—	4/6/90	1×10^{-5}	—	—	17,500
VA	3	1.2	—	—	5/14/90	1×10^{-5}	—	—	17,500
MN	2	0.00051	—	—	7/24/90	1×10^{-5}	30.0	276,000	—
NH	1	—	1.0	—	8/8/90	1×10^{-5}	—	—	17,500
TN(TEQ)[b]	2	—	1.0	—	1/17/91	1×10^{-5}	—	—	17,500
GA	5	—	1.2	—	1/23/91	1×10^{-5}	—	—	17,500
AL	10	1.2	1.2	—	2/20/91	1×10^{-5}	—	—	17,500
SC	3	—	1.2	—	3/13/91	1×10^{-5}	—	—	17,500
MS	3	1.0	1.0	—	3/28/91	1×10^{-5}	—	—	17,500
CA(TEQ)[b]	4	0.013	0.014	—	4/11/91	—	—	—	—
TX(TEQ)[b]	5	1.0	0.7	—	6/12/91	1×10^{-5}	10 or 15	—	13,900
LA	4	0.71	0.72	—	10/20/91	1×10^{-5}	20.0	—	9,700
AR	4	1.0	—	—	11/7/91	1×10^{-5}	7.5	—	17,500
AZ	1	—	0.003	0.2	1/10/92	—	—	—	—
AK	2	0.13	0.14	—	12/22/92	1×10^{-5}	—	—	—
FL	6	0.013	0.014	—	12/22/92	—	—	—	—
ID	1	0.013	0.014	—	12/22/92	—	—	—	—
MI	3	0.13	0.14	—	12/22/92	1×10^{-5}	—	—	—
WA	11	0.013	0.014	—	12/22/92	—	—	—	—
KY	2	0.013	0.014	—	—	—	—	—	—

a. New York state standard based on recommended limit of 10 ppt in fish tissue.
b. TEQ indicates standard set on the basis of total equivalents.
na = Not available.
Source: EPA, 1993a.

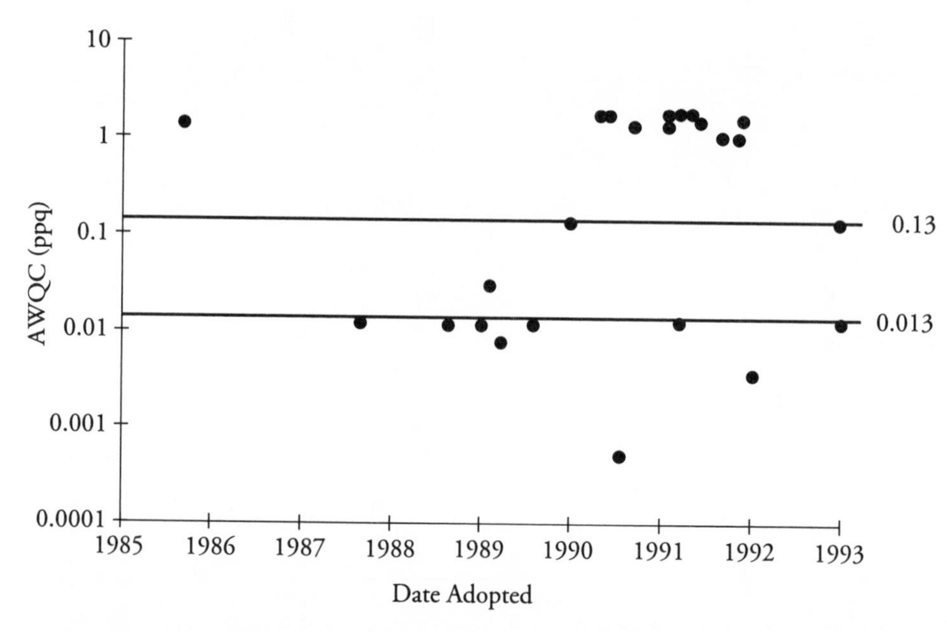

FIGURE 7.3 | Strictest state numerical ambient water quality criteria, and date adopted.
Source: EPA, 1993.

1989, Georgia adopted a temporary standard of 7.2 ppq (500 times higher than the original standard), which had been proposed by pulp and paper mills in the state. Two factors motivated the change: 1) advice from the state attorney general that the previous standard could not withstand the legal challenge threatened by industry, and 2) advice from a well-known public health official brought in by the industry, who indicated that a standard between 7.2 and 11 ppq would provide adequate safety and could be justified based on current research (Wade, 1993). Environmental groups were outraged by this development, and the local newspapers portrayed the change as a response to industry pressure. On March 27, 1990, the EPA notified Georgia that it would not approve the 7.2 ppq standard and would implement proceedings to impose its recommended criteria. Nonetheless, the next day Georgia formally adopted the 7.2 ppq standard and it removed all five of its mills from its 304(1) list, since they were in compliance with the new standard.

At the same time that Georgia was debating its standard and anticipating EPA disapproval, other southern states were also adopting standards. Maryland and Virginia relied on the FDA's cancer potency factor of (instead of the EPA's) and an acceptable risk level of 1×10^{-5} to establish a standard of 1.2 ppq. These states had clearly considered both the economic costs of a higher standard to the pulp and paper industries and the amount of flexibility allowed by the uncertainty in the risk assessment (Houck, 1991). Despite protests by environmentalists, the EPA approved the Maryland and Virginia standards, since they were within the range of scientific defensibility. Following the EPA's approval of the Maryland and Virginia standards and anticipating the agency's disapproval of its own 7.2 ppq standard, Georgia adopted and the EPA later approved a standard of 1.2 ppq.

In striking contrast to the relatively permissive standards adopted by most of the southern states, Minnesota adopted a standard that was 100 times stricter than the EPA's criteria. Minnesota used a bioconcentration factor (BCF) approximately fifty times higher than the EPA's BCF, and a fish consumption rate approximately five times higher, to arrive at a standard of 0.00051 ppq, which is orders of magnitude below the detection limit in fish tissue. In general, the dioxin limits selected by states appear to reflect the relative importance of the pulp and paper industry within the state, although notable exceptions like Oregon and Maine exist, where fishing and tourism are countervailing commercial interests.

The strictness of a state's dioxin limit further depends in part on how the state interprets the standards when issuing permits to specific mills. Since the standards apply to the bodies of water into which mills discharge and not directly to mill effluent, the amount of dilution that occurs is taken into account when establishing effluent limits for permits. However, states use different methods to account for dilution, with some basing estimates on a variation of mean flow (to estimate average concentrations), and others using a variation of low flow (to estimate more worst-case concentrations). In addition, since all the standards fall below the minimum level of detection for dioxin, states could decide to consider a mill in compliance with the standard so long as its effluent did not contain detectable levels of dioxin (that is, the level of detection could provide a "safe harbor"). However, states did not necessarily accept the "safe harbor" when issuing permits (some required mills to report detection limits or half of the detection limits in permits). Consequently mills fought to raise water quality standards above levels of detection to eliminate the ambiguity that could arise if a mill could not demonstrate compliance with the state standard because one could not detect dioxin concentrations that low.

An example of the regional importance in permit decisions occurred in the Columbia River basin in Oregon, Washington, and Idaho. These states each adopted a standard of 0.013 ppq, and at their request, in June 1990, the EPA established its first total maximum daily load (TMDL) as directed under CWA Section 303 for the eight mills discharging into the basin. While the mills initially opposed the standards, asserting that they were impossible to achieve and would create a competitive disadvantage with other U.S. mills, when the EPA finalized its proposal in February 1991, it indicated that much of the dioxin reduction required had already occurred, due to changes in manufacturing processes (Houck, 1991).

Integrated Quantitative Risk Assessment Results

While states were dealing with water quality issues, the EPA was coordinating the large number of risk assessments that were performed to satisfy the requirements of the 1988 consent decree. Figure 7.4 provides estimates of reasonable worst-case individual cancer risks that exceeded 1 in 1 million (1×10^{-6}) as given in the summary of the 1990 integrated risk assessment (EPA, 1990a). Although the risks are all presented in Figure 7.4, they are for different exposure groups and they rely on a number of different assumptions. The work group charged with preparing the EPA's integrated assessment agreed to use the risk results generated by different agencies and offices instead of attempting to reconcile risk assessment assumptions. The work group further agreed to report cancer risk estimates using each agency's cancer potency factor. Consequently, although each risk assessment generated a point estimate of risk, results of the integrated assessment given in Figure 7.4 show a range for each risk that reflects the order of magnitude of difference between the FDA's (least stringent) and the EPA's (most stringent) cancer potency factors. Ranges larger than this order of magnitude reflect differences in 1) exposure estimation approaches (for example, the EPA's surface water modeling versus the FDA's fish tissue monitoring to estimate fish concentrations of dioxin ingested by fishers), 2) mill-specific variability (such as differences in effluent

Source	Pathway	Exposed individuals	Upper bound individual lifetime cancer risk ($\geq 10^{-1}$ 10^{-2} 10^{-3} 10^{-4} 10^{-5} 10^{-6})
Effluent discharge	Ingestion of fish (EPA site-specific)	Average persons,	
		sport fishers,	
		subsistence fishers	
	Ingestion of fish (FDA generic)	Sport fishers,	
		subsistence fishers	
	Ingestion of water	Individuals near mills	
Sludge disposal (occupational)	Dermal contact with sludge or sludge-amended soil	Pulp mill water treatment wkrs.,	
		land disposal wkrs.	
	Inhalation of particles	Land disposal wkrs.	
	Inhalation of vapors	Land disposal wkrs.	
Sludge disposal (ambient)	Dermal contact with sludge-amended soil	Gardeners/subsistence farmers	
	Ingestion of sludge-amended soil	Gardeners/subsistence farmers	
	Ingestion of food produced on sludge-amended soil	Gardeners/subsistence farmers	
	Inhalation of particles	Individuals near land appl. sites	
	Inhalation of vapors	Individuals near land appl. sites	
	Ingestion of fish contaminated by runoff (from land disposal and land application sites)	Individuals near sites	
	Ingestion of water contaminated by runoff (from land disposal and land application sites)	Individuals near sites	
Pulp/paper manufacture (occupational)	Inhalation of particles	Paper mill wkrs.,	
		paper converting wkrs.,	
		nonwoven operations wkrs.	
	Dermal contact with pulp/paper	Pulp mill wkrs.	
Paper-food contact	Ingestion of food	Users of ice cream cartons,	
		users of frozen meals,	
		users of soup cups,	
		users of paper plates	

FIGURE 7.4 | Individual risk estimates above 1 in 1 million associated with releases from mills.
Source: EPA, 1990a.

concentrations and consequent water body concentrations for ingestion of fish and water), 3) job-specific variability (for instance, the differences in work duties leading to different rates of contact with pulp or paper in manufacturing), and 4) uncertainty (for example, regarding the use of protective equipment by people contacting sludge, or the amount of fish consumed).

In the assessment, the EPA and the FDA considered toxic equivalents because mill emissions included approximately five to ten times as much furan as dioxin and because both agencies assumed that furan was one-tenth as toxic as dioxin (EPA, 1989a). Consistent with the characteristics of mill releases, both the EPA and the FDA assumed that only dioxin and furan were of toxicological significance, and consequently they ignored the small contributions to TEQ that could be made by other dioxins and furans. The Consumer Product Safety Commission used only the results for dioxin in its assessments.

The integrated assessment identified four risks of "concern": those to 1) "humans from food packaging," 2) "terrestrial and avian organisms from land-spreading of pulp and paper sludge," 3) "humans and aquatic organisms from effluent," and 4) "humans and wildlife from land-applied pulp and paper mill sludge runoff" (EPA, 1990a: xix). The assessment also identified risks to workers and subsistence farmers contacting sludge as risks of possible concern. Although population risks were not estimated explicitly, the integrated assessment indicated that more than one case of cancer per year could be associated with some of the populations shown in Figure 7.4. For example, more than one case could occur for populations exposed to effluent discharges, since the individual risks predicted could be very high, and the same is true for populations exposed to food packaging and cellulose derivatives, since the exposure could occur for tens of millions of individuals.

While the industry performed its own risk assessments (see, for example, NCASI 1987a, b, c; NCASI 1988b, c, d), the results from these would not appear in Figure 7.4 because they did not produce estimates that exceeded a risk of 1 in 1 million (which the EPA implicitly used as a threshold for significant risk). The industry risk assessments tended to characterize risks in terms of the amount of dioxin that could be present that would not pose a significant risk (that is, by identifying the dioxin concentration associated with a risk of 1 in 1 million) and to use slightly different assumptions than those used by the EPA. Nonetheless, risk estimates can be inferred from these studies, and in a few cases (such as risks associated with using bleached paper coffee filters for heavy coffee drinkers), the risk estimates might have reached 1×10^{-6} if the industry had chosen to use the EPA's most stringent cancer potency factor instead of the FDA's factor.

The integrated assessment also provided some estimates of noncancer effects on humans and ecosystem effects associated with dioxin emissions from mills (not included in Figure 7.4). The integrated risk assessment indicated that effluent discharges from 27 percent of mills could potentially lead to toxic liver effects, that discharges from more mills could potentially cause reproductive effects from low-level and long-term exposure, and that subsistence fishers could be potentially at risk for reproductive effects (EPA, 1990a). These estimates are heavily dependent on the fish bioconcentration factor that is chosen in the risk equation. The integrated assessment also predicted that dioxin water concentrations immediately downstream of more than 80 percent of mills would exceed levels that are toxic to some aquatic species. It further suggested that some terrestrial species might experience reproductive effects from the contact with dioxin and furan in land-applied sludge. Thus, the results of the integrated risk assessment provided ammunition for the advocacy groups concerned about dioxin, although this was weakened by the fact that the EPA used outdated information about the industry, since it had already implemented significant process changes.

| Fish Alert

The risk estimates associated with consuming fish contaminated by dioxin in effluent were much higher than those associated with consuming contaminated water (Figure 7.4). Following release of the *National Dioxin Study* report (EPA, 1987a), which suggested that dioxin levels in fish had reached potential levels of concern, the EPA initiated a national study of chemical residues in fish to evaluate the presence of dioxin, dioxin-like compounds, and other toxic pollutants accumulating in fish (EPA, 1992). Dioxin samples in the study were collected between 1984 and 1989 (the majority in 1987), and the results were released in a 1989 draft report (EPA, 1989c). The study found a statistically significant relationship between pulp and paper mills using chlorine and fish tissue concentrations of dioxin and furan. This finding led to the conclusion that mills were a dominant source of these pollutants.

Only the FDA used the available fish tissue data generated by the national study to estimate risks for the 1990 integrated risk assessment. The FDA performed a "generic" risk assessment by using the available fish tissue data to estimate an average TEQ fish concentration for use as the point estimate in the risk assessment (FDA, 1990a). In contrast to this approach, the EPA estimated the risks for each mill in its risk assessment (EPA, 1990b) by using models and the mill-specific data from the 104-Mill Study to predict water and fish concentrations. The EPA elected not to use the National Bioaccumulation Study (EPA, 1989c) results in its assessment, primarily because data were unavailable for some mills (EPA, 1990b), although it could have used them to validate its model estimates. As a result of its modeling approach, the EPA generated a wide range of risk estimates for effluent discharge, reflecting the variability associated with different mills (shown in Figure 7.4).

The availability of fish tissue data led state environmental and health departments to issue fish consumption advisories or bans intended to protect the general public and sport and subsistence anglers from high risks associated with eating locally caught contaminated fish. In the 1990s, more than 1,000 water bodies have been under some type of advisory, with mercury, polychlorinated biphenyls, and pesticides accounting for approximately 90 percent of the total advisories. In 1994, 54 of the 1,533 existing advisories were for dioxin, and 18 of these in thirteen states applied to water bodies downstream of mills (AET, 1995).

In spite of the relatively small percentage of the total related to pulp and paper mills, these advisories have had a major impact on the industry. The total numbers of dioxin fish advisories associated with pulp and paper mills issued, rescinded, and in effect at the end of each year between 1983 and 1995 are shown in Table 7.3. Overall, approximately 30 percent of the 104 mills have been associated with an advisory at one time or another. The number of advisories in effect increased while the industry was becoming aware of and assessing the problem, and as the availability of mill-specific effluent data collected

TABLE 7.3 | Numbers of fish advisories for dioxin downstream of pulp and paper mills issued, rescinded, and in effect at the end of each year

YEAR	ADVISORIES ISSUED	NUMBER RESCINDED	TOTAL IN EFFECT
1983	1	0	1
1984	0	0	1
1985	1	0	2
1986	0	0	2
1987	3	0	5
1988	2	0	7
1989	6	0	13
1990	17	0	30
1991	4	4	30
1992	0	6	24
1993	0	3	21
1994	1	3	19
1995	0	1	18

Source: AET, 1995.

during the 104-Mill Study (NCASI, 1990a) facilitated the identification of mills as possible sources.

The large jump in the number of advisories in 1990 occurred largely in response to the EPA's mill-specific risk assessment (EPA, 1990b) and a press release it issued on September 24 that summarized the results (EPA, 1990c). The press release focused attention on twenty mills for which the estimated reasonable worst-case risk of cancer exceeded 1 in 10,000 (1×10^{-4}) for an average consumer eating fish caught downstream of the mill (EPA, 1990c). It further identified which mills had fish advisories in downstream water bodies and suggested that "states consider imposing fish consumption advisories or start site-specific monitoring programs" for the twenty highlighted mills if they were not already in effect (EPA, 1990c).

Placing the spotlight on the industry, the information in the press release was picked up by a number of newspapers, with the *San Francisco Chronicle* publishing the chart shown in Figure 7.5 (SFC, 1990). In its statements about the press release, the industry pointed out that the EPA's conclusions were based on old data that were collected prior to the implementation of process changes by many mills

PAPER MILLS CITED FOR DIOXINS

These mills were cited by the Environmental Protection Agency for posing high dioxin risks. Plants ranked by cancer risk; federal fish advisory status included:

Company/location	Cancer Risk	Fish consumption advisory
International Paper Co., *Georgetown, S.C.*	2 in 100	Yes
Union Camp Corp., *Franklin, Va.*	2 in 1,000	No
Buckeye Cellulose, *Perry, Fla.*	2 in 1,000	No
Weyerhauser Co., *Plymouth, N.C.*	2 in 1,000	Yes
Westvaco Corp., *Covington, Va.*	1 in 1,000	Yes
Georgia Pacific Corp., *Palatka, Fla.*	6 in 10,000	No
International Paper Co., *Moss Point, Miss.*	3 in 10,000	Yes
Temple-Eastex Inc., *Evadale, Texas*	3 in 10,000	No
Champion International, *Canton, N.C.*	2 in 10,000	Yes
Georgia-Pacific Corp., *Crosset, Ark.*	2 in 10,000	No
International Paper Co., *Texarkana, Texas*	2 in 10,000	No
International Paper Co., *Jay, Maine*	1 in 10,000	Yes
Boise Cascade Co., *Rumford, Maine*	1 in 10,000	Yes
St. Joe Paper Co., *Port St. Joe, Fla.*	1 in 10,000	No
Boise Cascade Co., *Deridder, La.*	1 in 10,000	No
Simpson Paper Co., *Anderson, Calif.*, and Louisiana Pacific Co., *Samoa, Calif.*	1 in 10,000	Yes
Simpson Paper Co., *Fairhaven, Calif.*	1 in 10,000	No
Weyerhauser Co., *Cosmopolis, Wash.*	1 in 10,000	No
Weyerhauser Co., *Everett, Wash.*	1 in 10,000	No
Leaf River Forest, *New Augusta, Miss.*	1 in 10,000	Yes

FIGURE 7.5 | Mills highlighted in EPA press release on dioxin-contaminated fish.

Source: Clarence Johnson, "Fish Near Paper Mills Called Tainted," *San Francisco Chronicle*, September 25, 1990 (SFC, 1990).

that had reduced dioxin emissions substantially (*Pulp and Paper,* 1990). The recognition of these process changes by states (discussed below) is evident in the downward trend in the total number of fish advisories for dioxin downstream of mills since 1991 (Table 7.3).

| Potential Liability

The presence of dioxin in mill effluent also opened the door to potentially costly litigation and damage payments to plaintiffs. Plaintiffs claimed fear of health harm, nuisance, and damage to property values. In some cases, risk assessments (mainly qualitative) were used to support the plaintiffs' cases.

For example, more than 8,000 plaintiffs filed suit in southern Mississippi against Georgia-Pacific's Leaf River Mill — ironically one of the most modern U.S. mills at the time. In November 1990, jurors in Mississippi awarded over $1 million for trespass, nuisance, loss of property value, and punitive damages to a homeowner residing on the Leaf River, miles downstream of the mill (Varchaver, 1993). Georgia-Pacific settled with a group of plaintiffs in a separate suit for an undisclosed amount in 1991, and was assessed over $3 million in punitive, nuisance, and emotional distress damages for other plaintiffs in a case that went to trial in 1992 (Varchaver, 1993). Both verdicts were reversed by the Mississippi Supreme Court in 1995, and a 1993 jury trial against Georgia-Pacific resulted in a verdict for the defense. Nonetheless, the threat of liability clearly provided more incentives for the industry to emphasize scientific uncertainty about the carcinogenicity and other toxic endpoints of dioxin, and to support scientific research investigating the mechanisms by which dioxin acts. It also may have provided mills with added incentives to ensure that they were in compliance with permits, so that they could state in court that they were never in violation of the state's health-based standard (*Pulp and Paper,* 1991: 31).

| Risks from Pulp and the Use of Paper Products

Motivated by the discovery of dioxin in mill effluent by the National Dioxin Study (EPA, 1987a), the EPA Office of Water began to explore whether consumers of paper products were exposed to dioxin. The office circulated a draft report that considered the risks from food and body contact with dioxin-contaminated paper (EPA, 1987b). The assessment used hypothetical assumptions of exposure to generate estimates of individual risks from contact with seven everyday paper products. The highest estimate of 1 in 10,000 (1×10^{-4}) came from using one paper coffee filter every day for fifty years, with the dioxin concentration of 10 ppt being 100 percent bioavailable and with 50 percent of the ingested dioxin being absorbed by the coffee drinker. Risks from paper plates, disposable diapers, tampons, and paper towels ranged from 1.4 to 6.6 in 1 million (1.4×10^{-6} to 6.6×10^{-6}), while risks from dinner napkins and uncoated paper were below 1 in 1 million.

This risk assessment, which was later characterized by the EPA (1990a) as a "scoping study," was highly criticized by both industry and EPA reviewers. Dr. Renate Kimbrough, a key EPA reviewer who had been responsible for preparing the Times Beach Superfund site risk assessment (Kimbrough et al., 1984), said in a memo to the assistant administrator that the draft was "not helpful and perhaps should not have been done" (EPA, 1987c). In particular, Dr. Kimbrough indicated that better information about the concentration and bioavailability of dioxin in pulp products and the extent to which dioxin migrates out of paper and is absorbed by humans should be obtained before EPA performed any other risk assessments.

Research within the industry was under way to obtain these key pieces of information. In April 1987, industry representatives met with the FDA to discuss analytical protocols and research needs. The industry performed a number of studies and issued its own risk assessment progress reports in 1987 (NCASI, 1987c, d)

In October 1988 the Canadian Government released results of a survey showing that five of eight whole milk samples from milk cartons made from bleached paper contained measurable levels of dioxin and furan (FDA, 1994). These results motivated Greenpeace to petition the U.S. Department of Agriculture to prepare an environmental impact statement associated with the portions of its school lunch program that provided funding for the purchase of milk in chlorine-bleached milk cartons, including the effects of disposal of the cartons. Not surprisingly, the USDA sought advice from the FDA, which was planning its own monitoring program.

In 1989, the FDA performed its own measurements for all five

U.S. manufacturers of pulp for milk cartons. Based on a study that compared dioxin measurements of packaged and unpackaged milk, the FDA found that low levels of dioxin were leaching into milk from the cartons (with four of fifteen samples being slightly above detection limits). When the FDA released the results, the *New York Times* reported that the FDA had set a three-year time limit for mills to implement process changes that would reduce levels of dioxin in pulp. The FDA indicated that the milk was "safe to drink" during the short time period required to make changes, and all five manufacturers promised to make the changes within months (NYT, 1989). Dr. Ellen Silbergeld of the Environmental Defense Fund represented the sentiment of the environmental community when she suggested that milk producers shift to alternative packaging (such as plastic containers) until manufacturers finished process changes (NYT, 1989).

Following the milk study, the FDA requested that the industry perform additional migration studies for those foods in contact with paper products that posed the greatest potential for exposure. The industry provided its research results to regulators, and the FDA used them in the development of risk estimates which were incorporated into the integrated risk assessment (EPA, 1990a). The integrated assessment estimated the risks associated with numerous pulp products and exposure pathways under different regulatory jurisdictions, as shown in Table 7.4. Even using conservative but reasonable assumptions, the assessment predicted negligible risks (below 1 in 10 million) for all pulp product and exposure pathways, except for those resulting from ingestion. Upper-bound estimates of risk associated with ingestion of foods in contact with pulp-based packaging were as high as 1 in 100,000 (1×10^{-5}) (see Figure 7.4), although important uncertainties remained regarding dioxin migration rates and food intake rates specific to food-contacting pulp products. In addition, a few of the upper-bound estimates of risk associated with ingestion of foods containing cellulose derivatives just reached 1 in 1 million (1×10^{-6}).

In accordance with the 1988 consent decree, the EPA elected to refer action on food-contact paper risks to the FDA and did so formally on December 26, 1990 (EPA, 1990d). Both the EPA and the FDA considered the risk estimates to be relatively low. Nonetheless, the EPA considered the risks to be "unreasonable" in accordance with section 9(a) of the Toxic Substances Control Act (EPA, 1990d), since the

TABLE 7.4 | Pulp-related risks considered in the integrated risk assessment under different agency jurisdictions

ROW	EXPOSURE PATHWAY	AGENCY WITH JURISDICTION	PULP-PRODUCTS
1	Dermal (from use of and body contact with products—both dry and liquid-mediated)	CPSC	Disposable diapers, paper towels, facial and toilet tissue, paper napkins, communication paper (uncoated bond, books, newsprint, magazines)
2	Dermal (from use of and body contact with products—both dry and liquid-mediated)	FDA	Sanitary pads, tampons, alcohol and other medical wipes, surgical apparel, medical absorbent fiber, examination gowns, disposable bedding, other medical devices
3	Dermal (from use of and body contact with cosmetic products containing cellulose derivatives—both dry and liquid-mediated)	FDA	Dentifrice, lotion, shampoo, wet cleansing wipes for adults and infants
4	Ingestion (of food contacting or packaged in pulp products)	FDA	Milk, cream, and juice cartons, coffee filters, coffee and soup cups, prepared meals on dual-oven trays, prepared meals on paper plates, popcorn microwave bags, donuts and sweet rolls in bakery cartons, ice cream and other frozen dairy cartons, tea bags, margarine wrap
5	Ingestion (of food and drug products containing cellulose derivatives used for anti-caking, thickening, and stabilizing)	FDA	High-fiber bread, tablet binders, laxitives, other foods (baked goods, candy, dairy products, sausage casings, dried fruits, diet beverages, and flavorings)
6	Inhalation and ingestion (from volatilization out of municipal landfills and from drinking groundwater contaminated by landfill leachate)	EPA	Waste paper

Source: EPA, 1990a.

industry had demonstrated its ability to reduce dioxin and furan levels, and all of the mills producing bleached paper for food-contact products had committed to a voluntary program a month earlier that would reduce levels of dioxin below 2 ppt in all products.

At the time of the referral, the FDA was deciding whether immediate regulatory action was necessary to address risks while the industry adopted its voluntary process changes. Based on a minor modification of its earlier risk assessment, the FDA determined that risk levels were low enough that continued use would be acceptable during the time needed for mills to implement process changes and for the FDA to perform a reassessment. The FDA performed monitoring studies, which verified that dioxin levels in products were below the 2 ppt limit, and also performed an assessment that estimated an upper-bound limit of individual lifetime cancer risk of 3 in 10 million (3×10^{-7}) for the 2 ppt limit (FDA, 1993). In April 1994 the FDA issued a notice indicating that it was still monitoring and evaluating risks, but it tentatively concluded that risks associated with food-contact paper products did not require further action (FDA, 1994).

| Dioxin-Contaminated Sludge

Sludge is produced as contaminants are removed during the waste water treatment processes that most mills adopted to meet biological oxygen demand and other effluent requirements established under the Clean Water Act. It results from sedimentation of wood fiber, lime, and other solids out of waste water and from the biological treatment of waste water. In 1990 the EPA characterized the disposal methods of the approximately 2,500,000 dry metric tons (DMTs) of sludge generated annually by the 104 mills producing bleached pulp (Table 7.5). The integrated risk assessment predicted insignificant risks (individual and population) from the 12 percent of sludge that was incinerated (EPA, 1990a). In contrast, some of the highest individual risk estimates in the integrated risk assessment (Figure 7.4) were associated with dioxin from the 88 percent of sludge which was placed on land in some capacity.

Similar to sludge generated by municipalities and other industrial processes, sludge from pulp and paper mills may be used beneficially to increase the organic content and nutrients in soil. The benefits of the sludge depend on its characteristics and the characteristics of the soil to

TABLE 7.5 | Disposal of sludge

DISPOSAL METHOD	QUANTITY (DMT)	PERCENT OF TOTAL QUANTITY	NO. MILLS (SOME REPORTING MULTIPLE METHODS)	COST PER DMT($)
Landfill	1,100,000	44	59	
75% on site				114
25% in municipal landfills				162
Surface impoundment	600,000	24	20	95
Land applications	300,000	12	7	
80% forest (silvicultural) land				76
10% agricultural land				40
10% reclaimed mines				31
Incineration	300,000	12	21	140
Distribution and marketing	200,000	8	7	92
Total	2,500,000	100	104	

Sources: EPA, 1990a; EPA, 1991b.

which it is applied. Land disposal methods differ in the extent to which they take advantage of benefits to soil from the sludge, and in cost. Table 7.5 provides cost estimates of the different disposal methods for an average-size mill. Costs may differ substantially for some mills due to the lack of availability of some options (for example, long distances to mine sites or other suitable land application sites). In general, landfills and surface impoundments, which contain the sludge and derive no benefits, cost twice as much per DMT as land applications and distribution and marketing, disposal methods that use the sludge as a lower-cost fertilizer replacement or supplement.

The integrated risk assessment predicted theoretically high risks associated with land disposal of sludge. The highest risk estimates resulted from ingestion of food grown and animals grazed on sludge-amended lands and from ingestion of fish contaminated by runoff from sludge-amended land into surface water. The risk assessment made a number of worst-case assumptions to characterize risks for a maximally exposed individual (MEI). For example, the EPA assumed that 1) no daily or final cover would be placed on landfills or surface impoundments and no runoff controls would be in place, 2) the site size would be large and the drainage area small, 3) the runoff concentration from the entire drainage area would be representative of the contaminant concentration in surface water with no further dilution, 4) grazing animals ingested sludge and soil as 8 percent of their diet, 5) subsistence farmers ingested food at the maximum daily consumption rates, and 6) the MEI consumed fish at a rate of 140 grams/day (EPA, 1990e).

In accordance with the consent decree, the EPA agreed to propose a rule under the Toxic Substances Control Act to address these risks by April 30, 1991. The TSCA requires the agency to reduce an "unreasonable" risk to human health or the environment, based on consideration of benefits and costs. On May 10, 1991, the EPA proposed a rule regarding the land application (including distribution and marketing) of sludge (EPA, 1991b) that included soil concentration limits, sludge management practices, and requirements for monitoring, record keeping, and periodic reporting.

In response to comments received from industry, the EPA significantly revised its discussion of sludge disposal practices and its human health and wildlife risk assessments in support of the proposed rule (EPA, 1991c, d). While the exposure and risk models remained unchanged, a number of the values of key parameters were made more realistic and less conservative. For example, in the revised assessment the EPA assumed that landfills and surface impoundments would be properly managed and would have seepage and runoff controls. Consequently, the risks from landfills and surface impoundments became insignificant, and the EPA did not consider further regulation of sludge going to landfills or surface impoundments.

Even after the EPA revised its risk assessment for land application, it found that the risks were still high enough to warrant regulation. In the revised assessment, the predicted MEI risks were still on the order of 1 in 1,000 (1×10^{-3}) and population risks ranged from 0.3 to 0.002 cases of cancer per year (EPA, 1991b). To develop a range of potential regulatory options, the EPA reversed the risk equation to estimate combinations of soil concentration and application-area size limits, which would lower risks for the MEIs to within the range of 1×10^{-4} to 1×10^{-6}.

The EPA performed a complete cost-benefit analysis for the regulatory options, and proposed to adopt a soil concentration limit of 10 ppt after weighing the scientific evidence for human health risks against the economic consequences of land application restrictions. The agency assumed that the 10 ppt soil concentration limit would limit risks to subsistence farmers to 1×10^{-4}. In the proposed rule, the EPA estimated that the cost of the nationwide 10 ppt limit would be approximately $5.5 million per year and that lower limits would produce virtually no incremental human health risk reduction, at comparatively high incremental costs. The proposal also considered imposing an additional application-area size limit of 1,000 hectares per drainage area in order to limit risks to subsistence fishers to 1×10^{-4}, and it indicated that choosing a risk level of 1×10^{-5} would effectively ban land application. The EPA considered the costs of imposing area limits to be unjustified, given the uncertainties inherent in the risk assessment for fishers.

Although the proposed soil concentration limit addressed high human risks, it left some predicted risks to wildlife above insignificant levels. In its analysis, the EPA used models to estimate that a soil concentration limit of 0.03 ppt would reduce exposures of the most exposed species considered (shrew and woodcock embryo) to an insignificant level. However, the proposal indicated that the limit was thirty-three times lower than the analytical limit of detection and would effectively ban land application, which the EPA did not advocate (EPA, 1991b). Instead, the EPA indicated that the proposed limitations were designed to force paper mills to develop all reasonable means of reducing dioxin and furan in sludge (EPA, 1991b), and not to eliminate potential benefits from appropriate land applications. Since the risk estimates were based on the sludge measurements from the 104-Mill Study, reductions of dioxin and furan concentrations in sludge that had occurred as a result of process changes were not considered explicitly. Nonetheless, the effects of these changes are mentioned in the rule, and they may have been a factor in the EPA's decision not to advocate a land application ban.

In November 1992, the EPA and the American Forest & Paper Association (AF&PA), an industry trade association, agreed to develop an environmental stewardship program for the practice of applying sludge to land. Following this agreement, the EPA informed plaintiffs party to the consent decree that it would delay finalizing the proposed sludge rule pending the 1995 promulgation of an integrated rulemaking for mills that the EPA expected to propose in late 1993 (described in the next section). Each paper mill participating in the environmental stewardship program would sign a memorandum of understanding with the EPA that would establish a maximum sludge concentration of 50 ppt TEQ, a maximum post-application soil concentration of 10 ppt TEQ (the same level that the EPA proposed), provisions for site management, and record keeping and monitoring requirements. Mills in all but the three states that had developed their own requirements (Maine, Wisconsin, and Ohio) were eligible to participate in the program.

In 1986, following measurements in Wisconsin conducted for the National Dioxin Study, mills in that state voluntarily stopped land application of sludge pending an agreement with state agencies. At the request of a Nekoosa Paper mill (prior to its purchase by Georgia-Pacific), the state and industry agreed to carry out a research program to assess the ecosystem effects of land application of sludge to forests. After several years of research, the program found that robin eggs were accumulating dioxin from the sludge. Based on these results, the state determined a mass limit that would be protective of robins (Thiel et al., 1995), although no mills in Wisconsin are currently applying sludge to forest lands (Goodman, 1996).

Wisconsin also considered the risks associated with application of sludge to farm land. In 1988 the Departments of Health and Natural Resources collaborated on a risk assessment that predicted high risks for a hypothetical MEI who ingested fruit and vegetables grown as well as meat and dairy products from cattle grazed on sludge-amended land, as well as local fish and game. To validate claims by the industry that no such MEIs existed, the Department of Natural Resources conducted a survey of approximately 400 people who lived near fields that had been used for mill-sludge land applications. The 1989 survey identified 183 full-time farmers, 1 of whom conformed reasonably well to the MEI exposure scenario, producing all of his own food at home (Nelson, 1996). The survey also identified two respondents who produced almost all of their own food and spent some time working for the local mill, a combination that was not considered in the risk assessment (Nelson, 1996). Based on these results, the state issued final guidelines for the soil concentrations in cattle grazing fields (0.5 ppt TEQ) and nongrazing fields (1.2 ppt TEQ) (Baker, 1993). These limits, as well as monitoring requirements, are incorporated into the permits of the one mill that provides its sludge for agricultural spreading (although the sludge does not contain detectable levels of dioxin or furan).

Both Maine and Ohio regulated dioxin in sludge based on risk assessments similar to Wisconsin's agricultural assessment (Keenan et al., 1988). In Maine, the state limited sludge concentrations to 250 ppt TEQ and final soil levels to 27 ppt TEQ, and restricted land application to areas not used to produce foods (plant and animal products) for human consumption (ME DEP, 1986). In Ohio, where sludge was used for mine reclamation, the state limited sludge concentrations to 100 ppt TEQ for one-time applications to soil (OH DEP, 1987).

| Pollution Prevention by the Industry

The industry has undertaken process changes since the 104-Mill Study that have dramatically reduced dioxins and furans generated and released to the environment. In 1988, mills began voluntarily switching from chlorine to chlorine dioxide in the first bleaching stage, upgrading bleaching plants, and implementing other process changes to reduce the formation of dioxins and furans (such as improving brownstock washing, using oxygen for delignification or to enhance extraction stages, and eliminating defoamers identified as dioxin precursors). Figure 7.6 shows how the dioxin concentration distributions for the industry have changed since the 104-Mill Study (NCASI, 1995). For each type of discharge (sludge, pulp, and effluent), the figure shows the fiftieth percentile (median), ninetieth percentile, and the maximum concentration measurements reported during each year, and the minimum levels of detection. The consistent decline of all of the statistics indicates that mills were not simply shifting the dioxins and furans from one media to another. Figure 7.7 shows how specific process changes led to an overall decline in dioxin effluent concentration for a single mill.

Dioxin pollution prevention by the industry as a whole has been significant. Figure 7.8 uses estimates of dioxin releases from each mill to estimate the total mass (kg) of dioxin and furan (TEQ) released by the industry since 1988 (NCASI, 1995). The figure shows that the industry reduced its TEQ releases between 1988 and 1990 (while the 104-

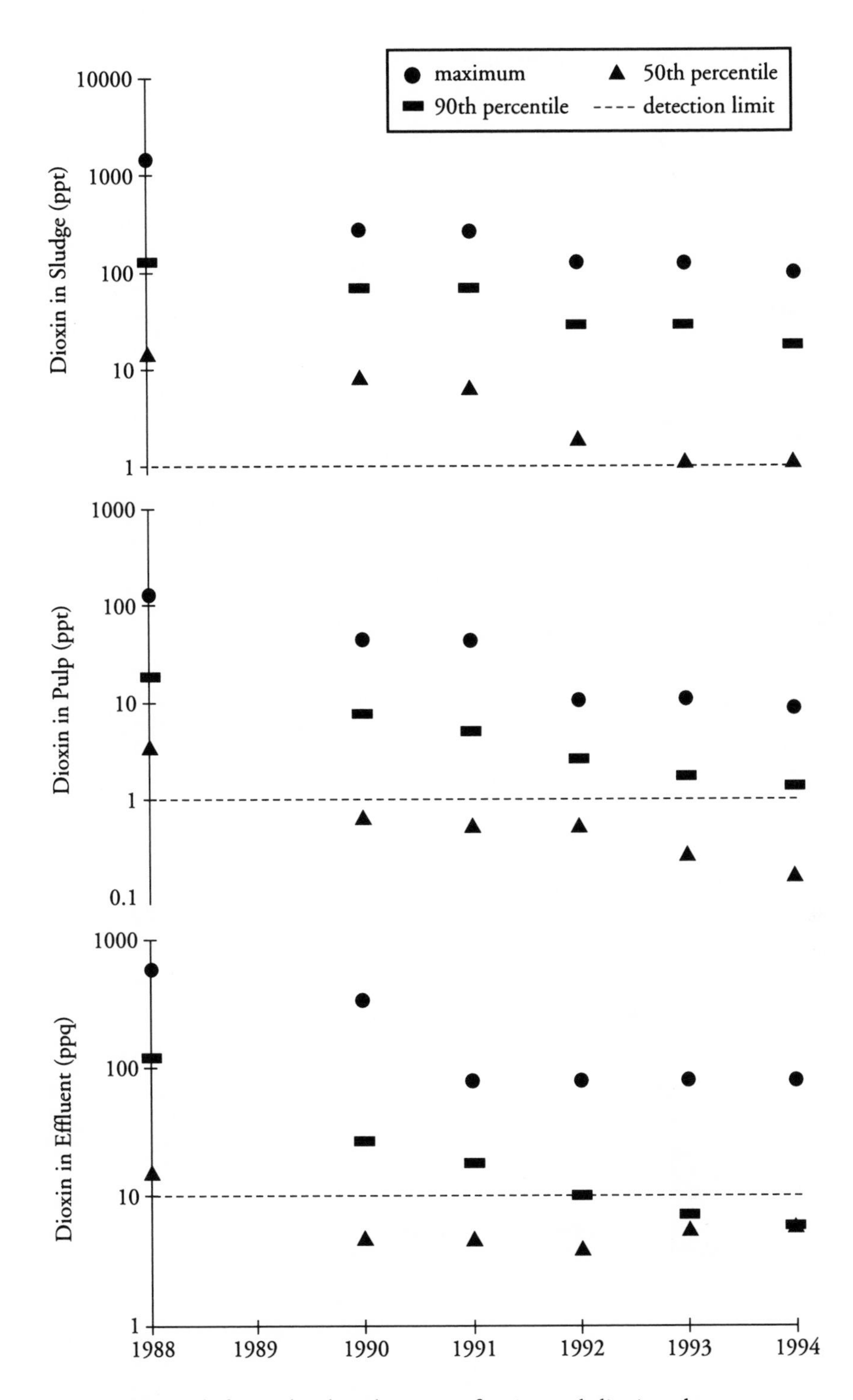

FIGURE 7.6 | Shifts in the distributions of estimated dioxin releases.
Source: NCASI, 1995.

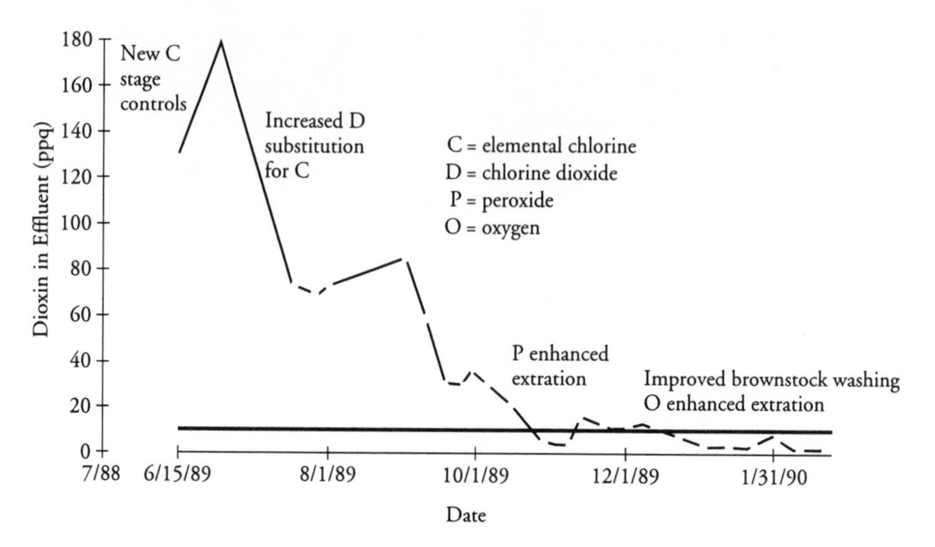

FIGURE 7.7 | Decline in effluent dioxin measurements for a single facility.
Sources: Potlatch, 1990; EPA, 1990f.

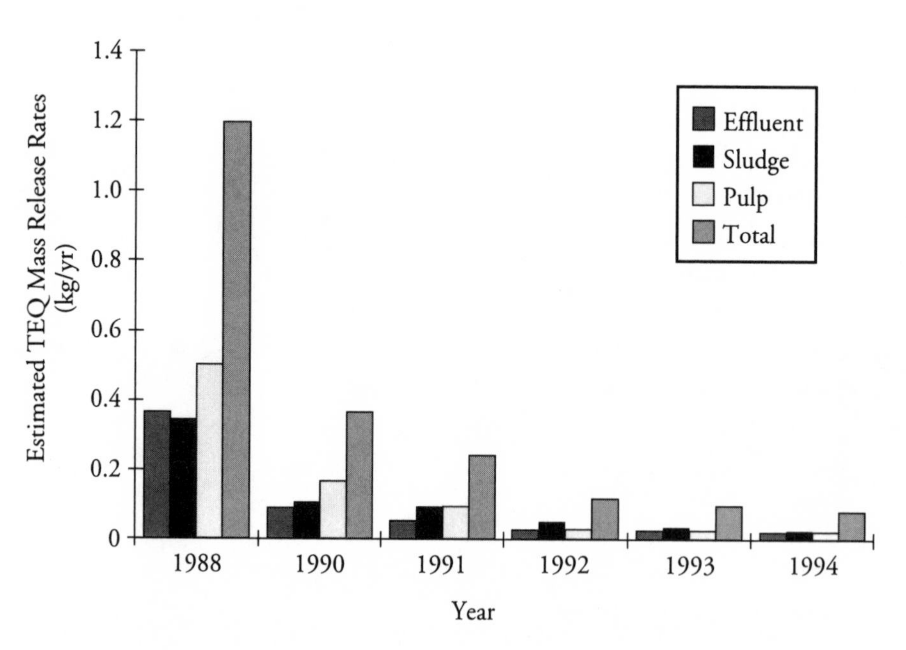

FIGURE 7.8 | Estimated total mass of toxic equivalents released by industry between 1988 and 1994.
Source: NCASI, 1995.

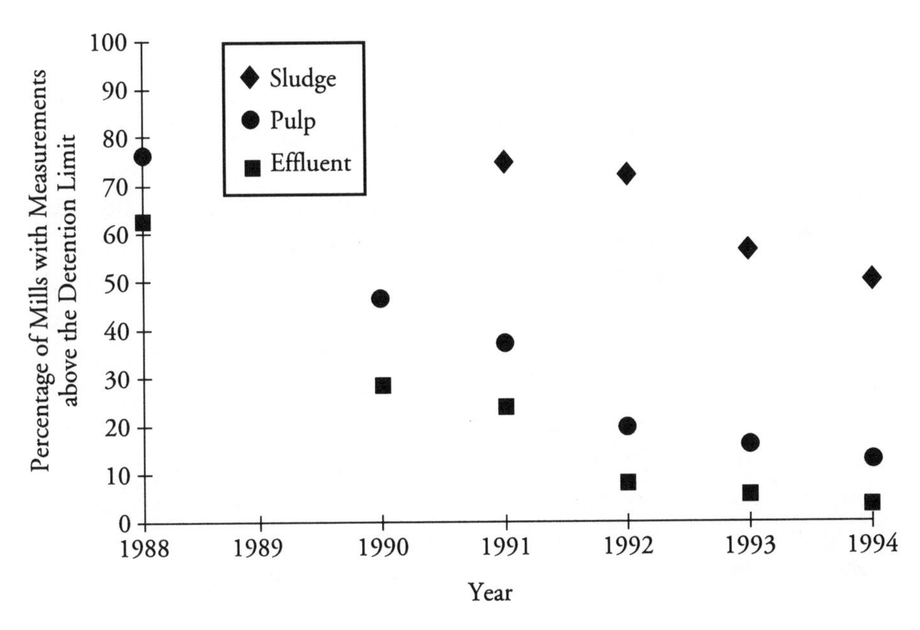

FIGURE 7.9 | Decreases in the number of mills reporting detectable levels of dioxin.
Source: Gillespie, 1996.

Mill and other intensive studies were being conducted). Figure 7.9 shows the decrease in the percentage of mills reporting concentrations above the minimum levels of detection (Gillespie, 1996). As of 1994, the industry had spent more than a billion dollars to implement measures to reduce the total mass of dioxin and furan (TEQ) emitted by the industry approximately 94% since 1988 (Cooper, 1993; NCASI, 1995). In its 1994 dioxin reassessment, the EPA estimated that the industry produced less than 3 percent of the dioxin generated in the United States, with incinerators accounting for more than 95 percent (EPA, 1994) (see Chapter 6 for discussion of the incineration case). A recent industry review also demonstrated declines in fish tissue dioxin concentrations downstream of mills (Abbott and Hinton, 1996), and the number of dioxin fish advisories for water bodies downstream of mills continues to decline (Table 7.3).

| Tougher Requirements?

Although the industry has made substantial progress reducing dioxin, in 1993 the EPA proposed new regulations for mills in an integrated rulemaking known as the "Cluster Rule" (EPA, 1993b). The rule responded to congressional demands for reductions of hazardous air pollutants under the Clean Air Act and for toxic substance control under the Clean Water Act (for all substances, not just dioxins and furans). The EPA had also intended to act on the 1991 proposed dioxin sludge rule, but indicated in the Cluster Rule proposal that it would be reevaluating sludge once the effects of water and air regulations were better understood. Prior to release of the proposal, industry was very supportive of an integrated rulemaking process because it would allow for coordinated compliance. However, when the EPA's proposed rule came out, the industry did not support the agency's choice of control technologies.

At the root of the controversy was the long-standing issue of whether mills should be required to invest in technology that would produce fewer or no chlorinated by-products, as some mills in Europe have done. While regulatory efforts in Canada and Europe have focused largely on the wide array of chlorinated organic compounds, efforts in the United States have centered more on dioxin and furan. For example, the discovery of dioxin strengthened concerns of the Swedish Environmental Protection Agency that highly persistent and toxic compounds were being emitted from mills and supported the SEPA's action to reduce total organic chlorine (TOCl) emitted from pulp and paper mills beginning in 1986 (Michel and Lagergren, 1991). In its regulation, Sweden chose to rely on a relatively inexpensive and reliable measurement technique that measures adsorbable organic halides (AOX) to estimate TOCl and establish its standard, because essentially all of the halides emitted from pulp and paper mills are chlorinated compounds. To lower AOX, mills typically need to replace from elemental chlorine with chlorine dioxide (using an elemental chlorine-free, or ECF, process), use oxygen delignification, or adopt a totally chlorine-free, or TCF, process. In the mid-1980s, Swedish mills began using ECF and TCF bleaching processes, and environmentalists argued that U.S. mills should follow suit.

In the proposed Cluster Rule, the EPA sought to establish a

standard for AOX of 0.156 kg/air-dried ton (ADT) of pulp, which is three times lower than the strictest current international standard (the Swedish standard is 0.5 kg/ADT; Finnish, 2.0 kg/ADT; Canadian, 0.8 kg/ADT). To meet this standard, mills would be required to install oxygen delignification technology and to switch completely from chlorine to chlorine dioxide for bleaching. However, the EPA and the industry disagreed significantly about the costs and benefits of the different regulatory options considered in the rule. In particular, the EPA estimated that the annual costs of the entire rule would be approximately $950 million, while its estimates of annual benefits ranged from $160 to $987 million (EPA, 1993b). In its own estimates, industry projected annual costs would exceed $2 billion and estimated annual benefits of between $8 and $46 million (Cooper, 1995). The EPA's monetized estimates of the benefits considered reductions in cancer and noncancer health effects from consumption of dioxin- and furan-contaminated fish in addition to the effects from lifting dioxin-related fish advisories. Overall, dioxin-related benefits appear to account for approximately one-third of all monetized benefits.

In its 1994 response to the proposed rule, the industry said that it would willingly commit to complete elimination of elemental chlorine for bleaching (that is, to become ECF), but it strongly objected to any requirements for the installation of oxygen delignification. The critical issue appears to be whether the EPA can demonstrate that having a low AOX standard will reduce environmental risks enough to justify the large expenditures that will be required by the industry to meet it. This is complicated by the nature of the AOX measurement, since it does not consider the relative toxicity, degradability, tendency to accumulate, or persistence of the different chlorinated organic compounds, and consequently it is difficult to use in a quantitative risk assessment.

In 1996, the EPA elected to address various components of the Cluster Rule separately, and thus far the issue of the benefits of reducing AOX remains unresolved. The EPA issued a notice in July 1996 regarding its goals for environmental improvement for the industry and describing its preliminary analysis of wastewater standards (EPA, 1996). The notice indicated that the EPA is now focusing on two options as the basis of establishing effluent guidelines and standards. For option A, which entails complete substitution of chlorine dioxide for elemental chlorine and implies less stringent for AOX requirements, the EPA estimated implementation costs of $140 million per year (in $1995). For option B, which entails oxygen delignification or extended cooking (to remove lignin) combined with complete substitution and implies more stringent AOX requirements, the EPA estimates that it will cost $155 million per year (in $1995). The EPA did not revise its estimates of benefits from those presented in 1993, but it suggested that the incremental environmental benefits of Option B over Option A included "reduced chronic toxicity to some aquatic life species" which may reflect "an incremental reduction in the potential for formation of dioxin and furan, which at many mills are no longer measurable by current analytical methods at the end-of-pipe, and a reduction in mass loadings of all chlorinated compounds which can be measured by the bulk analytical parameter [AOX]" (EPA, 1993b: 36838).

In its comments on the notice, the AF&PA asserted that the benefits of the two options would be the same, in spite of the large differences in cost (AF&PA, 1996). The AF&PA estimated the benefits from the implementation of either Option A or B would range from $5.9 to $27.5 million per year, and it reiterated its objections to the EPA's assertion that reductions in AOX would provide environmental benefits. The AF&PA estimated implementation costs would be $221 to $238 million per year for Option A, and $452 to $502 million per year for Option B. Based on its estimates, the AF&PA asserted that the costs of the regulation dwarf its benefits and that the EPA should make every effort to choose options that bring the costs closer to benefits (AF&PA, 1996).

| The Role of Risk Assessment

Risk assessment has played and continues to play a critical role in the reduction of dioxin and related compounds from the pulp and paper industry. In the case of pulp and paper mills, the importance of dioxin risk assessments was clear even before dioxin was discovered in mill releases. Since dioxin had gained the reputation as the "most toxic compound" from previous toxicity tests, the discovery of low concentrations of dioxin in mill discharges was enough to motivate the industry to spend millions of dollars to seek technological changes that would reduce or eliminate the problem.

The industry clearly moved quickly in its efforts to deal with dioxin and furan, generally committing to act voluntarily prior to the issuance of a regulation. The regulatory requirements imposed by the 1988 consent decree may have played a major role in the decisions to implement process changes at pulp and paper mills. First, it placed time limits for the EPA to provide guidance to states and to implement provisions of the 1987 amendments of the Clean Water Act. Thus the industry faced the certainty of mill-specific limits on the discharge of dioxin into surface waters. Second, by placing a definite time limit on completion of a comprehensive multimedia risk assessment, the consent decree focused attention on all types of emissions at the same time. Comparisons of the risks from different media showed risks of concern associated with each of the three types of releases (effluent, pulp, and sludge). Consequently, mills became subject to a flurry of potential regulatory activity from state and federal agencies (not to mention adverse publicity, consumer concerns, and tort litigation), which meant that the sooner mills could eliminate dioxin and furan the better. In addition, the multimedia approach of the risk assessment also foreclosed the possibility of shifting the dioxin and furan to a different medium (in particular, from effluent to sludge).

The limits of science and technology influenced the ability of regulators to enforce risk-based standards and the choices of control strategies used by mills. The fact that it was difficult to detect dioxin analytically in mill effluent was important, because it meant that enforcement agencies had to address the issue of how to deal with nondetectable amounts of dioxin. Overall, the industry has sought to lower its emissions of dioxin in all media below minimum levels of detection mainly by going elemental chlorine–free, but it has not opted to pursue more costly totally chlorine–free technologies that could further reduce its releases of other chlorinated substances. The fact that it is difficult to measure dioxin and furan in mill effluent complicates attempts to develop correlations between them and AOX.

The industry, which has a long history of environmental regulation and technological change, took a cooperative position with regulators in the case of dioxin at the same time that it contested some of the government's scientific determinations. This cooperation with regulators and participation in the risk assessment process allowed the industry to have more input in regulatory decisions. For example, the risk assessments that industry performed to support more permissive water quality standards in the South may have given some mills more time to evaluate alternative technologies. In doing its own assessments, the industry was able to challenge the assumptions made by different agencies, and to sponsor and perform research aimed at resolving scientific uncertainties. Risk and cost-benefit considerations did work to discourage regulators from mandating what some environmentalists continue to want: a ban on the use of any substances containing a chlorine molecule in the paper bleaching process.

Although it is not clear what motivated mills to take action precisely when they did, some mills may have been influenced by consumer concerns. Consumer concern about dioxin in milk and other food products led to a quick response by mills that manufacture pulp for food-contact products, which then set an important regulatory precedent for the rest of the industry by establishing that technology existed to reduce dioxin levels below detection.

Overall, the relatively high theoretical risk estimates associated with fish consumption, and consequent fish advisories, provided strong incentives for industry to adopt process changes and reduce dioxin levels in effluent quickly, especially with high-risk estimates for specific companies being highlighted in the popular press (Figure 7.5). Process changes were further required as mills had their effluent permits evaluated, although in some cases quantitative risk assessments were used to argue for water quality criteria that mills could meet without process changes (particularly in the South). Risk assessment results for sludge motivated the EPA and states to consider implementing restrictions on land applications of sludge, with risks to wildlife playing a major role in the evaluation of regulatory options. Because of the benefits of land application, the EPA and states did not advocate outright bans but instead sought to compel industry to explore means of reducing dioxin and furan in sludge. Since mills faced regulatory pressure for both effluent and sludge, and consumer concerns about dioxin in pulp, the industry had to find a way to reduce or eliminate its inadvertent production of dioxin and furan.

Noncancer and ecological risk assessments for this case appear to have played a very limited role with respect to regulatory actions and decision making. In the case of sludge, the EPA was unwilling to base its proposed soil concentration limit on the possible risks to wildlife. In spite of documented noncancer effects in wildlife, cancer risks have dominated the risk assessment and regulatory efforts — presumably because they are more "significant" at the risk levels of concern (see Figure 7.2).

The industry considers its dioxin reduction to be a major success, along with its cooperative risk-based approach to regulation (Cooper, 1993). Whether the industry will view the Cluster Rule similarly depends on what the rule (or separate rules) ends up requiring. Unlike some European mills, which have opted for a TCF bleaching sequence, mills in the United States have preferred to switch to ECF processes. One U.S. mill, which responded to the dioxin problem by going totally chlorine free now, exports its pulp because the pulp is considered to be less bright than other market pulp. Brightness appears to be an important attribute in paper, at least for American consumers.

Environmentalists also can take pride in much of the progress that has been made. A recent report by the Finnish Environment Agency gives them credit for motivating much of the development of TCF processes and their implementation in Sweden (FEA, 1996).

Currently, little research exists regarding the environmental impacts of TCF processes, although the Finnish Environment Agency concluded that "according to the studies to date neither ECF nor TCF bleaching concepts invariably produces effluents with a lower toxic potency than the other. . . . The stringent demands on reduced AOX-emissions from the Swedish pulp industry may therefore be considered to have been set by 'precautionary reasons' and not as a consequence of scientifically established damage caused by chlorinated organic substances occurring in receiving waters" (FEA, 1996: abstract and p. 7). In this country, the controversy over elemental chlorine–free versus totally chlorine–free paper production remains. Some environmentalists are not yet satisfied with the U.S. industry's choice to go ECF but not TCF, and they fear that difficulty in doing quantitative risk assessments for adsorbable organic halids hinders efforts to induce the industry to eliminate chlorine entirely.

| Sources

ABC, 1983. "How Dangerous Is Dioxin?" *ABC News Nightline* Transcript for March 15, 1983, show #482 (New York: American Broadcasting Companies, Inc.).

Abbott, J. D., and S. W. Hinton, 1996. "Trends in 2,3,7,8-TCDD Concentrations in Fish Tissues Downstream of Pulp Mills Bleaching with Chlorine," *Environmental Toxicology and Chemistry,* 15(7):1163–1165.

Allen, L. H., R. M. Berry, B. I. Fleming, et al., 1988. "Evidence that Oil-Based Additives Are a Potential Indirect Source of the TCDD and TCDF Produced in Kraft Bleach Plants," paper presented at the Eighth International Symposium on Chlorinated Dioxin and Related Compounds, Umea, Sweden, August.

Alliance for Environmental Technology (AET), 1995. *Eco-System Recovery: Liftings of Fish Consumption Advisories for Dioxin Downstream of U.S. Pulp Mills* (Washington, D.C.: Alliance for Environmental Technology, August). Report based on EPA national listing of fish consumption advisories and verification by state agencies.

American Forest & Paper Association (AF&PA), 1996. "Comments Submitted on Behalf of the American Forest & Paper Association With Respect to EPA's Notice of Availability: Effluent Limitations Guidelines, Pretreatment Standards, and New Source Performance Standards; Pulp, Paper, and Paperboard Category; National Emission Standards for Hazardous Air Pollutants for Source Category: Pulp and Paper Production," August 14 (Washington, D.C.: AF&PA).

Axegard, P., 1988. "Improvement of Bleach Plant Effluent by Cutting Back on Cl_2," *TAPPI Proceedings,* 1988 International Pulp Bleaching Conference, Orlando, June.

Baker, B., 1993. "Modification to WPDES Permits: Landspreading Guidelines for Sludge Contaminated with 2,3,7,8-Tetrachlorodibenzo-para-Dioxin and Furan," memorandum to Mary Jo Kopecky, Wisconsin Department of Natural Resources, April 23, 1993.

Banbury, 1991. M. A. Gallo, R. J. Scheuplein, and K. A. van der Heijden, eds., *Biological Basis for Risk Assessment of Dioxins,* Report #35 (Cold Spring Harbor, N.Y.: Cold Spring Harbor Laboratory Press).

Berry, R. M., B. I. Fleming, R. H. Voss, et al., 1989. "Toward Preventing the Formation of Dioxins during Chemical Pulp Bleaching," *Pulp & Paper Canada* 90(8):48–58.

Bertazzi, P. A., A. C. Pesatori, D. Consonni, et al., 1993. "Cancer Incidence in a Population Accidentally Exposed to 2,3,7,8-Tetrachlorodibenzo-para-Dioxin," *Epidemiology* 4(5):398–406.

Colburn, T., D. Dumanoski, and J. Myers, 1996. *Our Stolen Future: Are We Threatening Our Fertility, Intelligence, and Survival? A Scientific Detective Story* (New York: Dutton).

Consent Decree, 1988. *EDF and NWF v. Thomas,* Consent Decree, July 27, 1988, U.S. District Court Civil Action No. 85-0973).

Cooper, J. S., 1993. "Dioxin: An Industry Success Story," *TAPPI Journal* 76:10.

———, 1995. "Statement to the President's Commission on Risk Assessment and Risk Management," J. S. Cooper, Vice President, Environment and Regulatory Affairs, American Forest & Paper Association, January 11.

Courtney, K. D., D. W. Gaylor, M. D. Hogan, et al., 1970. "Teratogenic Evaluation of 2,4,5-T," *Science* 168:864–866.

Environment Ontario, 1988. *Kraft Mill Effluents in Ontario* (Toronto, Ontario: Ministry of the Environment). Report prepared by N. Bonsor, N. McCubbin, and J. B. Sprague, April.

Environmental Protection Agency (EPA), 1981. *Risk Assessment on (2,4,5-Trichlorophenoxy) Acetic Acid (2,4,5-T), (2,4,5-Trichlorophenoxy) Proprionic Acid, 2,3,7,8-Tetrachlorodibenzo-p-dioxin (TCDD),* Office of Health and Environmental Assessment, Carcinogen Assessment Group, EPA/-600/6-81-003, NTIS PB81-234825 (Washington, D.C.: Environmental Protection Agency).

———, 1983. *Dioxin Strategy,* Office of Water Regulations and Standards and the Office of Solid Waste and Emergency Response, EPA/600/6-91/002A (Washington, D.C.: Environmental Protection Agency).

———, 1984. *Ambient Water Quality Criteria for 2,3,7,8-Tetrachlorodibenzo-p-dioxin (TCDD),* Office of Water Regulations and Standards, EPA 440/5-84-007 (Washington, D.C.: Environmental Protection Agency).

———, 1985. *Health Assessment Document for Polychlorinated Dibenzo-p-Dioxins,* Final Report, Office of Health and Environmental Assessment, EPA/600/8-84/014F (Washington, D.C.: Environmental Protection Agency).

———, 1987a. *National Dioxin Study,* Office of Solid Waste and Emergency Response, EPA/530-SW-87-025 (Washington, D.C.: Environmental Protection Agency).

———, 1987b. *Exposure and Risk Assessment of Dioxin in Bleached Kraft Products,* Office of Water Regulations and Standards, Draft report prepared by Arthur D. Little, Inc. under Contract No. 68-01-6951 (Washington, D.C.: Environmental Protection Agency).

———, 1987c. Memorandum from Renate D. Kimbrough, Director for Health and Risk Capabilities, to John Moore, Assistant Administrator for Pesticides and Toxic Substances, September 28, as cited in *Environmental Reporter* 18 (October 16):1555–1566.

———, 1988a. *A Cancer Risk-Specific Dose Estimate for 2,3,7,8-TCDD,* External Review Draft and Appendicies, Office of Health and Environmental Assessment, EPA/600/6-88/007 Aa and EPA/600/6-88/007 Ab (Washington, D.C.: Environmental Protection Agency).

———, 1988b. "Interim Strategy for the Regulation of Pulp and Paper Mill Dioxin Discharges to Waters of the United States," memorandum from Rebecca W. Hanmer, EPA Acting Assistant Administrator for Water, to Water Management Division Directors and NPDES State Directors, August 9, 1988.

———, 1989a. *Interim Procedures for Estimating Risks Associated with Exposures to Mixtures of Chlorinated Dibenzo-p-dioxins and Dibenzofurans (CDDs and CDFs) and 1989 Update,* Risk Assessment Forum, EPA/625/3-89/016, NTIS PB90-145756 (Washington, D.C.: Environmental Protection Agency).

———, 1989b. "National Pollutant Discharge Elimination System; Surface Water Toxics Control Program." *Federal Register* 54:23868, June 2 (Washington, D.C.: Environmental Protection Agency).

———, 1989c. *Bioaccumulative Pollutants in Fish — A National Study,* Volume D and Appendicies A and B, Draft Report, Office of Water Regulation and Standards, December (Washington, D.C.: Environmental Protection Agency).

———, 1990a. *Integrated Risk Assessment for Dioxins and Furans from Chlorine Bleaching in Pulp and Paper Mills,* Office of Toxic Substances, EPA/ 560/5-90-011 (Washington, D.C.: Environmental Protection Agency).

———, 1990b. *Risk Assessment for 2,3,7,8-TCDD and 2,3,7,8-TCDF Contaminated Receiving Waters from U.S. Chlorine-Bleaching Pulp and Paper Mills,* Office of Water Regulation and Standards, August (Washington, D.C.: Environmental Protection Agency).

———, 1990c. "EPA Releases Risk Estimates for Eating Dioxin-Contaminated Fish," Press Office, Release R-158, September 24 (Washington, D.C.: Environmental Protection Agency).

———, 1990d. "Dibenzo-para-dioxins/Dibenzofurans in Bleached Wood Pulp and Paper Products; Referral for Action." *Federal Register* 55:53047, December 26 (Washington, D.C.: Environmental Protection Agency).

———, 1990e. *Assessment of Risks from Exposure of Humans, Terrestrial and Avian Wildlife, and Aquatic Life to Dioxins and Furans from Disposal and Use of Sludge from Bleached Kraft and Sulfite Pulp and Paper Mills,* Office of Solid Waste, EPA 560/5-90-013, July (Washington, D.C.: Environmental Protection Agency).

———, 1990f. *Summary of Technologies for the Control and Reduction of Chlorinated Organics from the Bleached Chemical Pulping Subcategories of the Pulp and Paper Industry,* Office of Water Regulation and Standards and Office of Water Enforcement and Permits, April 27 (Washington, D.C.: Environmental Protection Agency).

———, 1991a. "EPA's Scientific Reassessment of Dioxin: Background Document for Public Meeting on November 15, 1991" (Washington, D.C.: Environmental Protection Agency, October 1).

———, 1991b. "Proposed Regulation of Land Application of Sludge from Pulp and Paper Mills using Chlorine and Chlorine Derivative Bleaching Processes." *Federal Register* 56:21802, May 10 (Washington, D.C.: Environmental Protection Agency).

———, 1991c. "Human Health Risk Assessment for Dioxin in Sludge: Technical Support Document for the Proposed Land Application Rule," Office of Solid Waste, April (Washington, D.C.: Environmental Protection Agency). Revision of Chapter 6 of EPA, 1990e.

———, 1991d. "Environmental Risk Assessment for TCDD and TCDF-Contaminated Pulp Sludges on Terrestrial and Aquatic Wildlife," Office of

Pesticides and Toxic Substances, Health and Environmental Review Division, April (Washington, D.C.: Environmental Protection Agency). Revision of Chapter 6 of EPA, 1990e.

———, 1992. *National Study of Chemical Residues in Fish,* Office of Science and Technology, EPA 823-R-92-008a (Washington, D.C.: Environmental Protection Agency).

———, 1993a. *Tracking Report: State Water Quality Criteria for Dioxin (2,3,7,8-TCDD),* Office of Water Regulations and Office of Science and Technology Standards and Applied Science Division, September 22 (Washington, D.C.: Environmental Protection Agency).

———, 1993b. "Effluent Limitations Guidelines, Pretreatment Standards, and New Source Performance Standards; Pulp, Paper, and Paperboard Category; National Emission Standards for Hazardous Air Pollutants for Source Category: Pulp and Paper Production, Proposed Rule." *Federal Register* 58:66078, December 17 (Washington, D.C.: Environmental Protection Agency).

———, 1994. *Health Assessment Document for 2,3,7,8-Tetrachlorodibenzo-p-Dioxin (TCDD) and Related Compounds,* External Review Draft, three vols., Office of Health and Environmental Assessment, EPA/600/BP-92/001a, EPA/600/BP-92/001b, and EPA/600/BP-92/001c (Washington, D.C.: Environmental Protection Agency).

———, 1996. "Effluent Limitations Guidelines, Pretreatment Standards, and New Source Performance Standards; Pulp, Paper, and Paperboard Category; National Emission Standards for Hazardous Air Pollutants for Source Category: Pulp and Paper Production, Notice." *Federal Register* 61:36835, July 15 (Washington, D.C.: Environmental Protection Agency).

Fingerhut, M. A., W. E. Halperin, B. S. Marlow, et al., 1991. "Cancer Mortality in Workers Exposed to 2,3,7,8-Tetrachlorodibenzo-p-Dioxin," *New England Journal of Medicine* 342:212–218.

Finnish Environment Agency (FEA), 1996. The Aquatic Environmental Impact of Pulping and Bleaching Operations — An Overview, Publication Number 17 (Helsinki: Finnish Environment Agency, Oy Edita Ab).

Food and Drug Administration (FDA), 1990a. *Carcinogenic Risk Assessment for Dioxins and Furans in Fish Contaminated by Bleached-Paper Mills,* Report of the Quantitative Risk Assessment Committee, March 15 (Washington, D.C.: Food and Drug Administration).

———, 1990b. *Carcinogenic Risk Assessment for Dioxins and Furans in Fish Contaminated by Bleached-Paper Mills,* Report of the Quantitative Risk Assessment Committee, April 20 (Washington, D.C.: Food and Drug Administration).

———, 1990c. *Carcinogenic Risk Assessment for Dioxins and Furans in Cellulose Derivatives Used in Foods and Ingested Drug Products,* Report of the Quantitative Risk Assessment Committee, April 24 (Washington, D.C.: Food and Drug Administration).

———, 1993. *Upper-Bound Lifetime Carcinogenic Risks from Exposure to Dioxin*

Congeners from Foods Contacting Bleached Paper Products with Dioxin Levels Not Exceeding 2 ppt, Report of the Quantitative Risk Assessment Committee, January 27 (Washington, D.C.: Food and Drug Administration).

———, 1994. "Polychlorinated Dibenzo-p-dioxins and Polychlorinated Dibenzofurans in Bleached Food-Contact Paper Products; Response to Referral for Action by the Environmental Protection Agency and Request for Comment." *Federal Register* 59:17384, April 12 (Washington, D.C.: Food and Drug Administration).

Gillespie, W., 1996. Personal communication with William Gillespie, National Council of the Paper Industry for Air and Stream Improvement, April.

Goodman, B., 1996. Personal communication with Beth Goodman, Wisconsin Department of Natural Resources, February.

Hise, R. G., 1989. "Split Addition of Chlorine/pH Control for Reducing Formation of Dioxins," *TAPPI Proceedings,* 1989 International Pulping Conference, October, Seattle.

Houck, O. A., 1991. "The Regulation of Toxic Pollutants under the Clean Water Act," *Environmental Law Reporter* 21:10528–10560.

Keenan, R. E., M. M. Sauer, F. H. Lawrence, et al., 1989. "Examination of Potential Risks from Exposure to Dioxin in Sludge Used to Reclaim Abandoned Strip Mines," in D. J. Paustenbach, ed., *The Risk Assessment of Environmental and Human Health Hazards: A Textbook of Case Studies* (New York: John Wiley & Sons).

Kimbrough, R. D., H. Falk, P. Stehr, and G. Fries, 1984. "Health Implications of 2,3,7,8-Tetrachlorodibenzodioxin (TCDD) Contamination of Residential Soil," *Journal of Toxicology and Environmental Health* 14:47–93.

Kimmig, J., and K. H. Schulz 1957. "Berufliche akna (sog. chlorakne) durch chloriette aromatische zyklische ather" (Occupational Chloracne Caused by Aromatic Cyclic Ethers), *Dermatologia* 115:540–546.

Kleppe, P. J., and S. Storebraten, 1985. "Delignifying High-Yield Pulps with Oxygen and Alkali," *TAPPI Journal* 68(7):68–73.

Kociba, R. J., D. G. Keyes, J. E. Beyer, et al., 1978. "Results of a Two Year Chronic Toxicity and Oncogenicity Study of 2,3,7,8-Tetrachlorodibenzo-p-Dioxin in Rats," *Toxicology and Applied Pharmacology* 46:279–303.

Kringstad, K. P., and A. B. McKague, 1988. "Bleaching and the Environment," *TAPPI Proceedings,* 1988 International Pulp Bleaching Conference, Orlando, June. See also addendum with the same title by K.P. Kringstad, F. De Sousa, L. Johanson, et al.

Liebergott, M., B. van Lierop, B. C. Garner, and G. J. Kubes, 1983. "Bleaching a Softwood Kraft Pulp without Chlorine Compounds," *TAPPI Proceedings,* 1983 International Pulping Conference, October, Houston.

Luken, R. A., 1990. *Efficiency in Environmental Regulation: A Benefit-Cost Analysis of Alternative Approaches* (Boston: Kluwer Academic Publishers).

ME DEP, 1986. "Interim Standards for Sludge and Residuals Containing Polychlorinated Dibenzo-p-Dioxins and Polychlorinated Dibenzo-p-Furans (PCDDs and

PCDFs)," Part D, Amendment to Chapter 567, Rules for Land Application of Sludge and Residuals. Maine Department of Environmental Protection, Augusta.

Michel, I., and S. Lagergren, 1991. "The Swedish Policy for Regulating Chlorinated Organic Compounds Including Dioxin in the Pulp and Paper Industry," *Water Report* 1(2):4–6 (published by MYU K. K., Tokyo).

Miller, S. A., 1983. "Statement by Sanford A. Miller, Ph.D., Director, Bureau of Foods, Food and Drug Administration, Public Health Service, Department of Health and Human Services, before the Subcommittee on Natural Resources, Agriculture Research and Environment, Committee on Science and Technology, U.S. House of Representatives," June 30, 1983.

National Council of the Paper Industry for Air and Stream Improvement (NCASI), 1987a. *Assessment of Human Health Risks Related to Exposure to Dioxin from Land Application of Wastewater Sludge in Maine,* Technical Bulletin 525 (New York: National Council of the Paper Industry for Air and Stream Improvement, Inc.).

———, 1987b. *Land Treatment Effects on Wildlife Populations in Red Pine Plantations,* Technical Bulletin 526 (New York: National Council of the Paper Industry for Air and Stream Improvement, Inc.).

———, 1987c. *Assessment of Potential Health Risks from Dermal Exposure to Dioxin in Bleached Paper Products,* Technical Bulletin 534 (New York: National Council of the Paper Industry for Air and Stream Improvement, Inc.).

———, 1987d. *First Progress Report on the Assessment of Potential Health Risks from Use of Bleached Board and Paper Food Packaging and Food Contact Products,* Special Report 87-11 (New York: National Council of the Paper Industry for Air and Stream Improvement, Inc.).

———, 1988a. *U.S. Environmental Protection Agency/Paper Industry Cooperative Dioxin Screening Study,* Technical Bulletin 545 (New York: National Council of the Paper Industry for Air and Stream Improvement, Inc.). Also published by the EPA under the same title, Office of Water Regulations and Standards, EPA 440/1-88-025.

———, 1988b. *Risks Associated with Dioxin Exposure through Inhalation of Paper Dust in the Workplace,* Technical Bulletin 537 (New York: National Council of the Paper Industry for Air and Stream Improvement, Inc.).

———, 1988c. *Assessment of the Risks Associated with the Potential Exposure Dioxin through the Consumption of Coffee Brewed Using Bleached Paper Coffee Filters,* Technical Bulletin 546 (New York: National Council of the Paper Industry for Air and Stream Improvement, Inc.).

———, 1988d. *Assessment of Potential Health Risks to Pulp and Paper Mill Workers from Dermal Exposure to Dioxin in Bleached Pulp, Paper, and Paper-Based Products,* Technical Bulletin 549 (New York: National Council of the Paper Industry for Air and Stream Improvement, Inc.).

———, 1990a. *U.S. EPA/Paper Industry Cooperative Dioxin Screening Study: The 104-Mill Study,* Technical Bulletin 590 (New York: National Council of the Paper Industry for Air and Stream Improvement, Inc.).

———, 1990b. *An Intensive Study of the Formation and Distribution of 2,3,7,8-*

TCDD and 2,3,7,8-TCDF during the Bleaching of Kraft Pulps, Technical Bulletin 591 (New York: National Council of the Paper Industry for Air and Stream Improvement, Inc.).

———, 1992. *A Screening Study of the Treatability of Dioxins and Furans in Bleach Plant Filtrates and Mill Wastewaters,* Technical Bulletin 626, (New York: National Council of the Paper Industry for Air and Stream Improvement, Inc.).

———, 1995. *Progress in Reducing the TCDD/TCDF Content of Effluents, Pulps and Wastewater Treatment Sludges from the Manufacture of Bleached Chemical Pulp,* Special Report 95-12 (New York: National Council of the Paper Industry for Air and Stream Improvement, Inc.).

National Toxicology Program (NTP), 1982. *Bioassay of 2,3,7,8-Tetrachlorodibenzo-p-dioxin for Possible Carcinogenicity (Gavage Study),* Technical Report Series Number 102 (Research Triangle Park, N.C.: National Toxicology Program).

Nelson, E., 1996. Personal communication with Ed Nelson, Wisconsin Department of Natural Resources, February. Discussion of unpublished results of the Central Wisconsin Food Survey.

New York Times (NYT), 1974. David Burnham, "Scientist Urges Congress to Bar Any Use of Pesticide 2,4,5-T," *The New York Times,* August 10.

———, 1989. Philip J. Hilts, "Cartons Found Leaching Dioxin to Milk," *The New York Times,* September 1.

Office of Technology Assessment (OTA), 1989. *Technologies for Reducing Dioxin in the Manufacture of Bleached Wood Pulp,* U.S. Congress, Office of Technology Assessment, OTA-BO-O-54 (Washington, D.C.: Government Printing Office).

Ohanian, N. K., 1993. *The American Pulp and Paper Industry, 1900–1940: Mill Survival, Firm Structure, and Industry Relocation* (Westport, Conn.: Greenwood Press).

OH DEP, 1987. "Guidelines for Land Application of Paper Mill Sludge," Ohio Dept. of Environmental Protection, April 30, 1987, Columbus.

Peterson, R. E., H. M. Theobald, and G. L. Kimmel, 1993. "Developmental and Reproductive Toxicity of Dioxins and Related Compounds: Cross-Species Comparisons," *Critical Reviews in Toxicology* 23(3):283–335.

Pirkle, J. L., W. H. Wolfe, D. G. Patterson, et al., 1989. "Estimates of the Half-life of 2,3,7,8-TCDD in Vietnam Veterans of Operation Ranch Hand," *Journal of Toxicology and Environmental Health* 27:165–171.

Pohjanvirta, R., M. Unkila, and J. Tuomisto, 1993. "Comparitive Acute Lethality of 2,3,7,8-Tetrachlorodibenzo-p-Dioxin (TCDD), 1,2,3,7,8-Pentachlorodibenzo-p-Dioxin, and 1,2,3,4,7,8-Hexachlorodibenzo-p-Dioxin in the Most TCDD-Susceptible and the Most TCDD-Resistant Rat Strain," *Pharmacology and Toxicology* 73:52–56.

Poland, A., E. Glover, and A. S. Kende, 1976. "Stereospecific, High Affinity Binding of 2,3,7,8-Tetrachlorodibenzo-p-Dioxin by Hepatic Cytosol: Evidence that the Binding Species Is the Receptor for Induction of Aryl Hydrocarbon Hydrolase," *Journal of Biological Chemistry* 251:4936.

Portier, C., A. Tritscher, M. Kohn, et al., 1993. "Ligand/Receptor Binding for 2,3,7,8-TCDD: Implications for Risk Assessment," *Fundamental and Applied Toxicology* 20:48–56.

Pryke, D. C., and D. W. Reeve, 1985. "Chlorine Dioxide in the Chlorination Stage," *TAPPI Notes,* p. 49.

Pulp and Paper, 1990. "EPA Lists Mills with High Dioxin Levels," *Pulp and Paper* November: 31.

———, 1991. "IP, G-P Charged with Contaminating Waterways," *Pulp and Paper* February: 31.

Rao, M. S., V. Subbarao, J. D. Prasad, and D. G. Scarpelli, 1988. "Carcinogenicity of 2,3,7,8-Tetrachlorodibenzo-p-Dioxin in the Syrian Golden Hamster," *Carcinogenesis* 9:1677–1679.

Rapson, W. H. 1963. *The Bleaching of Pulp* (New York: Technical Association of the Pulp and Paper Industry).

Reeve, D. W. 1984. "E_O Installations Survey: June 1984," *TAPPI Journal* 67(10):110–111.

Roberts, L., 1991. "Dioxin Risks Revisited," *Science* 251:624–626.

San Francisco Chronicle (SFC), 1990. Clarence Johnson, "Fish Near Paper Mills Called Tainted: State Warns Against Eating Catch from Dioxin-Laden Waters." *The San Francisco Chronicle,* September 25.

Science Advisory Board (SAB), 1990. *Review of Draft Documents: A Cancer Risk Specific Dose Estimate for 2,3,7,8-TCDD and Estimating Exposures to 2,3,7,8-TCDD,* Science Advisory Board Ad Hoc Dioxin Panel, EPA-SAB-EC-90-003 (Washington, D.C.: Environmental Protection Agency).

———, 1995. *Re-evaluating Dioxin: Science Advisory Board's Review of EPA's Reassessment of Dioxin and Dioxin-like Compounds,* EPA-SAB-EC-95-021 (Washington, D.C.: Environmental Protection Agency).

Sodergren, A., B. E. Bengtsson, P. Jonsson, et al., 1988. "Summary of Results from the Swedish Project Environment/Cellulose," *Water Science Technology* 20(1):49–60.

Thiel, D. A., S. G. Martin, B. B. Goodman, and J. R. Sullivan, 1995. "Use of Loading Rates to Establish Dioxin Criteria for Land Application of Sludge," *Environmental Toxicology and Chemistry* 14(8):1443–1450.

Tritscher, A. M., J. A. Goldstein, C. J. Portier, et al., 1992. "Dose-Response Relationships for Chronic Exposure to 2,3,7,8-Tetrachlorodibenzo-p-Dioxin in a Rat Tumor Promotion Model: Quantification and Immunolocalization of CYP1A1 and CYP1A2 in the Liver," *Cancer Research* 52:3436–3442.

Van Strum, C., and P. Merrell, 1987. "No Margin of Safety: A Preliminary Report on Dioxin Pollution and the Need for Emergency Action in the Pulp and Paper Industry" (Washington, D.C.: Greenpeace, Inc.).

Van Miller, J. P., J. J. Lalich, and J. R. Allen, 1977. "Increased Incidence of Neoplasms in Rats Exposed to Low Levels of 2,3,7,8-Tetrachlorodibenzo-p-Dioxin," *Chemosphere* 6(10):537–544.

Varcharver, N., 1993. "Muddy Waters," *The American Lawyer* July/August:52–59.

Wade, R. M., 1993. "Regulating a Toxic Chemical: The Dioxin Controversy in Georgia," *Georgia State University Law Review* 9:717–736.

Weeks, L. H., 1916. *A History of Paper Manufacturing in the United States, 1690–1916* (New York: The Lockwood Trade Journal).

EIGHT

Risk Management: Green or Dirty?

John D. Graham
Jennifer Kassalow Hartwell

When industrial pollution is so severe that it is obvious to the ordinary citizen, the case for strong environmental regulation becomes overwhelming. If rivers are burning, the air smells, or drinking water causes acute illness, common sense causes citizens and their elected representatives to respond. Such overt symptoms of industrial pollution were sometimes evident — and at least memorable — to many Americans at the time of Earth Day 1970. And these overt symptoms of industrial pollution, coupled with tenacious environmental advocacy efforts, led to bipartisan decisions by the United States Congress to authorize strong regulatory programs under laws such as the Clean Air Act, the Clean Water Act, and the Safe Drinking Water Act.

The various federal environmental laws are complex, but they share a basic presumption that industrial pollution should be reduced because the costs to industry and consumers of reducing pollution will generally be worthwhile in light of resulting improvements in environmental quality. The Environmental Protection Agency's resulting regulations have often compelled industry to make significant reductions in pollution within specified deadlines, regardless of the precise magnitude of the risks of pollution or the benefits and costs of pollution control. For example, the commonly employed "technology-based" approach to environmental regulation was aimed at inducing all firms within an industry to approach (or exceed) the performance of the cleanest firms within the industry (McGarity, 1991). In response to this approach, many firms made substantial environmental progress, often reducing emissions to air and water by 70 to 95 percent, and even more in some cases. Although this approach has caused many inefficiencies, the overall estimates of benefits have often been greater than estimated costs (Freeman, 1993; Hahn and Hird, 1991).

As the turn of the century approaches, the residual problems of industrial pollution in the United States are less obvious to ordinary citizens (NAPA, 1995; Graham, 1995). Emissions to air and water certainly continue, but the amounts are sufficiently reduced in many locations that the case for further reductions may no longer be self-evident. In many cases, it is only through scientific methods that pollutants can be detected and their adverse effects predicted, and even these predictions have considerable uncertainty. Newly recognized environmental problems will not necessarily be targets of community outrage. For example, the potentially serious problems created by global warming, stratospheric ozone depletion, and indoor radon are invisible and cannot be discerned by citizen intuition (Cross, 1994).

At the same time, it is becoming more apparent than ever before that pollution control programs can be costly. Economists are fond of reminding us that, absent technological breakthroughs, the extra cost of reducing pollution by 95 percent instead of 90 percent may exceed the total cost of achieving the first 90 percent reduction in pollution. In fact, the EPA estimates that the United States as a whole is now investing $150 billion per year in pollution control, and this cost is increasing at a rate of more than 10 percent per year. This level of investment is larger, as a fraction of gross domestic product, than what is invested by most developed countries around the world (EPA, 1990b). States and localities are now joining industry in asking hard questions about whether more stringent environmental regulations of industrial pollution are worthwhile.

It seems clear that policy in the United States must evolve to reflect changing circumstances. If the residual amounts of industrial pollution (and their effects on health and ecology) are less evident to ordinary citizens, how will the EPA persuade the public and their elected representatives to renew their historical commitments to continuous progress against industrial pollution? If the costs of industrial pollution control are rapidly increasing, how will politicians resist pressures from industry — particularly industries facing intense global competition — to slow the pace of tightening regulations? The answers to these questions presumably rest in a more scientific approach to environmental policy, one that is less dependent on citizen outrage and more reliant on evidence about risk, benefit, and cost.

| The Risk Management Approach

According to William Ruckelshaus, who saw these dilemmas emerging when he returned to the helm of the EPA in 1983, a principled analytic framework was needed that weighs the social benefits from further pollution reductions against the incremental costs to industry and consumers that might be incurred (Ruckelshaus, 1983). Since the gains may not be fully discernable by common sense, he advocated the emerging science of "risk assessment." The costs of regulation would be more rigorously weighed in the context of risk management decisions informed by cost-benefit analysis and related tools. Ruckelshaus and his immediate successors (Lee Thomas and William Reilly) sought to persuade a distrustful Democratic Congress that future environmental policies should be based on risk management.

Environmental activists and their allies in Congress were adamantly against the Ruckelshaus approach from the moment he espoused it. For example, David Doniger of the Natural Resources Defense Council saw "risk management" as an unethical ploy to tolerate pollution (Doniger, 1989).

| The Case for Risk Management: Should We Be Persuaded?

This kind of hostility to the idea persists today, as is evident from the intense opposition to the risk-oriented provisions in the Republican "Contract with America" that passed in the House in March 1995 but were weakened and ultimately stalled in the Senate due to opposition from organized environmentalists and Vice President Albert Gore. Is this hostility warranted, or does it reflect the natural reaction of established powers to the force of changing circumstances and new ideas?

In the previous chapters of this book, we examined case studies of various industries in which the risk-management framework was attempted by the EPA, with or without encouragement from Congress. In this chapter, we offer some observations, based on the case studies, about whether risk management is a promising approach to further reducing pollution throughout American industry. We also consider whether the concerns of environmentalists are warranted and how environmental policies should be redesigned to reflect the legitimate concerns of those who are now defending the status quo in environmental policy.

| Morality or Pragmatism?

According to *The American Heritage Dictionary* (second edition), the first definition of the word "pollute" is "to render morally impure; corrupt." Since we should certainly aspire to communities that are morally pure and free of corruption, this definition leads naturally to the notion of zero pollution. An underlying philosophical problem with the risk-management framework is that it suggests, implicitly if not explicitly, that a little bit of moral impurity or corruption is okay if the tangible dangers of pollution are sufficiently small or the costs of prevention sufficiently large. Thus, we should never expect all environmentalists to be enthusiastic about risk management.

While the first dictionary definition of a word is generally considered preferable, a key to progress in future environmental policy may require closer attention to the second definition of "pollute": "to make unfit for or harmful to living things, especially by the addition of waste matter." An implication is that zero emissions may not be necessary if residual emissions are small enough to be "fit" for living things or free of "harm" to living things. Science presumably has an important role to play in determining what is harmful to living things, though moral judgments are also necessary to determine which effects on living things are harmful. In other words, judgments about what is pollution require a mix of scientific and moral judgments rather than exclusive reliance on either one.

In our industrial case studies, regulations seem to have been driven — at least on the surface — by the second definition of pollution. Emitting lead in gasoline exhaust was considered pollution because it harms children and property. Emitting perchloroethylene from dry cleaners was pollution because it may be toxic to workers and residents who inhale it. The fugitive emissions from coke plants were pollution not only because the odor is offensive but also because the fumes are known to cause cancer among heavily exposed workers. The dioxins emitted from municipal incinerators and paper mills were pollution because they persist and bioaccumulate in the environment and may ultimately cause adverse effects in humans and other species.

The closest we come to observing regulation driven by an exclusively moral definition of pollution is in the case of regulation concerning the role of chlorofluorocarbons in depleting the stratospheric ozone layer. Here, there was a tangible concern that increases in ultraviolet radiation (resulting from ozone depletion) could increase the incidence of skin cancer and other tangible effects on nonhuman as well as human species. But knowledge of these potential effects may not have been necessary to stimulate a global response. There was a subjective uneasiness about allowing chemical emissions from human activities to perturb the natural environment, on a global scale, with unknown consequences. Even in this case, however, the concern may ultimately have been rooted in tangible considerations. Perhaps this was a prudent reaction to the possibility that unforeseen catastrophic effects for humans and ecosystems might result from damaging the ozone layer.

The necessity for regulators to present at least qualitative evidence of potential danger is evident from the case study of dry cleaners. The industry fought bitterly the official carcinogen classification, because it perceived that this qualitative designation, by itself, would trigger regulatory responses (as well as potential problems in the marketplace and the courtroom). If perchloroethylene had tested negative in animal tests for carcinogenicity and had been universally accepted as having no meaningful role in smog formation, it is unlikely that any regulatory pressures for emission reduction would have been forthcoming. In other words, man-made chemical emissions into the environment may be of little or no concern if there is no reasonable basis for suspecting harm to humans or other living things.

In summary, the case studies suggest that there are moral concerns about man-made emissions of waste matter, but they are typically rooted in the notion that such emissions may harm living things. This is good news for the risk-management framework, since it suggests that actors in environmental policy are interested in what the emerging science of risk assessment can predict about the adverse effects of pollution on living things. If our case studies are an accurate guide, the predominant public interest in pollution prevention is pragmatic or tangible in nature.

| Must Risk Be Quantified Prior to Regulating?

Some reform advocates have argued that the probability and severity of the effects of pollution should be quantified before regulations are designed, implemented, and enforced. On this question, there appear to be some important distinctions revealed by the case studies.

Initial reductions in pollution from uncontrolled levels may be required based on qualitative risk assessment (assuming that the predicted health effects are severe), but regulations of residual amounts of pollution usually require quantitative risk assessments. Moreover, regulatory responses tend to be stricter when larger populations are exposed to risk and more lenient when smaller populations are exposed. However, efforts to persuade communities through quantitative risk assessment that hazardous technologies are "safe" face an uphill battle once qualitative dangers are suggested. We consider each of these generalizations below.

The cases suggest that qualitative indications of potential danger from pollution are often sufficient to motivate precautionary responses, even in the absence of quantification. The EPA's 1973 decision to reduce the lead content of gasoline was motivated, at least in part, by a qualitative concern for the health of children and adults. Once perc was determined to be a carcinogen, there were pressures for regulatory controls of the dry cleaning industry even though the EPA's quantitative risk assessment suggested that the magnitude of the elevated cancer risks to nearby residents and entire populations were small. Likewise, a qualitative understanding of the range of possible adverse effects of stratospheric ozone depletion was decisive in regulating CFCs, even though efforts were ultimately made to quantify the impacts on the worldwide incidence of skin cancer. In each of these cases, the initial steps toward pollution reduction were justified by *qualitative* risk assessment.

On the other hand, increasing the stringency of existing regulations on industry may be difficult to justify without quantitative indications of danger. Coke plants had been regulated for decades, based on qualitative health and quality-of-life concerns about particulate and sulphur pollution (Graham and Holtgrave, 1990). But it was the EPA's quantitative estimates of cancer risk to nearby residents — and the threat of their use in the 1990 legislation — that played an important role in motivating Congress and the industry to embrace tougher technology-based standards.

Likewise, the 1973 decision to phase out lead in gasoline was supported by qualitative health concerns (rather than quantitative risk assessment), but the 1985 decision by the EPA to accelerate the phaseout might not have been possible without the quantitative indications of economic and human health benefits. It is certainly difficult to imagine this regulation passing muster in the Reagan administration without a fairly rigorous accounting of benefits.

A common theme in all the cases is the potential of severe harm to people from pollutant exposures (supplemented in some cases by harm to nonhuman species). The human health effects of concern were *not* mild, temporary illnesses or discomforts but fatal, irreversible, and painful diseases such as cancer. Since less severe effects are likely to become more central in decisions about smaller pollutant concentrations, severity of harm may become a more salient consideration in future environmental regulations. If future regulations are based more on mild or subtle health effects instead of fatal ones, decision makers may be more inclined to insist on the quantitative estimates of frequency of harm (that are sometimes deemed to be unnecessary when severe harms are contemplated).

Harms that might be widespread in the human population, those of particular concern to public health scientists, appear to trigger stronger regulatory policies than those that affect small numbers of people. For example, the case studies of lead and chlorofluorocarbons, with huge populations at risk, witnessed regulatory bans, while regulators declined to shut down industries contributing to localized pollution problems (for example, coke producers and dry cleaners). Although environmentalists raised various "equity" arguments on behalf of nearby residents, these arguments — regardless of their merits — were ultimately not compelling enough to justify industrial shutdowns.

The case study of municipal incinerators reveals the difficulty of siting a new facility that has been tagged as "hazardous," even though the probability of health damage to nearby residents may not be very large. In Connecticut, some success in siting incinerators was achieved by comparing a proposed facility's incremental risk to a health-based standard, but it is not clear how generalizable this experience would be. Interestingly, the same degree of community hostility was not evident toward coke plants that had been operating for decades and posing much larger risks of disease to nearby residents. Whether it makes sense for a community to reject a modern incinerator while tolerating an old coke plant is a question worthy of future study by psychologists as well as economists and risk assessors.

| Is There a Role for Cost-Benefit Analysis?

In neoclassical microeconomics, the costs of pollution prevention are the "opportunity costs" of devoting scarce capital and labor to pollution prevention rather than to the production of other goods and services valued by consumers. This approach to cost estimation was used by the EPA in the cases of coke production, dry cleaning, and pulp and paper production. When material inputs are banned or phased out for environmental reasons, the standard approach to cost estimation is to estimate the increase in costs of final products (if any) attributable to the use of more costly substitute inputs. This approach to cost estimation was used by the EPA in the phasing out of leaded gasoline and chlorofluorocarbons.

There are certainly technical obstacles to making unbiased and precise estimates of the actual costs of environmental regulations. Some of the critical information needed to make cost estimates may be proprietary, forcing the regulatory analyst to make informed judgments and guesses. Once regulations are in place, firms often make creative technological or operating choices that induce costs that are lower than estimated *ex ante*. The "learning curve" with regulation may result in smaller long-run costs than are estimated prior to regulation. Conversely, the costs of operating new technologies or control equipment may prove greater than anticipated, particularly if the EPA relies too heavily on optimistic claims by the suppliers of pollution control technologies (Graham and Holtgrave, 1990).

Despite these obstacles to making valid and precise cost estimates, the EPA produced cost estimates in each case study, and yet these estimates (while definitely uncertain to various degrees) were not the focal point of controversy in any of the case studies. One possible reason that cost estimates are not disputed is that laws give limited or no weight to cost in decision-making criteria. However, our interpretation is that the degree of technical consensus about cost estimation is far greater than the degree of technical consensus about how baseline risks and estimates of risk reduction should be calculated in environmental problems. Nonetheless, it seems reasonable to demand that agency cost estimates be scrutinized for biases and uncertainties in the kinds of peer review processes that are often applied to risk assessments, especially if cost is to play a greater role under future environmental laws (Finkel and Golding, 1995).

The more serious conundrum revealed by the case studies is the lack of a rigorous approach to compare the estimated costs of environmental regulations with the expected reduction in estimated risks. When the costs of regulation are quantified in dollar units but the benefits are purely qualitative in nature (for example, less human exposure to perchloroethylene, a substance shown to cause cancer in animals at high doses), there is little guidance for (or constraint imposed on) regulatory discretion. Thus, in the 1980s the EPA and the states — with at least passive support from Congress — chose to pressure the dry cleaning industry into investing resources in technology to reduce perc emissions, even though the dangers of perc were understood only qualitatively. If more antiregulation policymakers had been in control of the Congress, the dry cleaning industry might have escaped regulation, since the qualitative benefits might have been judged to be inadequate.

Even when risks were quantified (as with the cancer risks of coke oven emissions), the EPA was sometimes reluctant to express these risks in dollar units, perhaps in deference to equity concerns for a relatively small number of residents living near coke plants. In the case of chlorofluorocarbons, the EPA analysts made some ballpark estimates of the monetary value of reducing ultraviolet radiation, but the decision to regulate seemed to be more rooted in a common sense judgment that the costs were small compared with the unknown and potentially large (yet unquantified) dangers of ozone depletion. The EPA's inability to quantify and monetize the full benefits of protecting the ozone layer may yet undermine the agency's phaseout of the more essential uses of chlorofluorocarbons, since it is possible that the Clean Air Act Amendments of 1990 and the Montreal Protocol's provisions will be reopened by a future Congress.

The reluctance to express costs and environmental benefits in a common unit (such as dollars) may be philosophical in nature. Some argue that cost-benefit analysis lacks an ethical underpinning. Assuming the desire to act morally, why should we refrain from it? (Kelman, 1981). The money saved or spent is irrelevant if there is no consideration given to right and wrong. However, cost-benefit analysis is governed by a sense of right. It is rooted in the utilitarian branch of philosophy, according to which action is taken so long as it maximizes net benefits (Leonard and Zeckhauser, 1986). Thus, the argument is not whether cost-benefit analysis is amoral, but rather which ethical school is worthy of adoption. Since the costs of environmental protection are of increasing concern and since "rights-oriented" approaches to environmental protection have yet to find a coherent role for cost information, it seems likely that the cost-benefit approach, warts and all, will assume greater importance in the future.

The EPA's decision to accelerate the phaseout of lead in gasoline offers the clearest case of a quantitative and monetized analysis driving regulatory decision making. And it is instructive that this analysis apparently led to a stricter regulatory outcome than would have occurred if only political or qualitative judgments had been allowed to dominate the deliberations of the EPA under President Reagan. This case study establishes clearly that cost-benefit analysis is a two-edged sword, rather than the one-sided tool that some naïve industrialists advocate and some naïve environmentalists oppose.

Given the current state of the art of environmental benefits analysis, it seems clear that a strictly quantitative and monetary approach to cost-benefit analysis of pollution control is not feasible. Despite the recent progress in risk assessment (Holland and Sielken, 1993; NRC, 1994) and environmental economics (Freeman, 1993; Tolley, Kenekel, and Fabian, 1994), our ability to quantify and monetize everything that seems important in environmental decision making is deficient. In the case of the lead phaseout, for example, numerous benefit categories were identified that could not be quantified.

A logical response to this predicament that has been proposed in legislative form by Senator Daniel Patrick Moynihan (D-N.Y.) is a long-term research program at the EPA aimed at enhancing our technical abilities to quantify risks, rank them, and express them in dollar units (Moynihan, 1993). Unfortunately, Moynihan's proposal never received serious support from environmentalists or industry, even though it represents a logical step toward making Ruckelshaus's framework a more workable guide to decision making.

Lacking the Moynihan plan, it now appears that agencies will be forced to learn through experience how to estimate environmental benefits under a mandate from Congress or the Office of Management and Budget, or both, a mandate that could apply to many more regulations than OMB orders now cover. For example, Congress has already mandated cost-benefit analysis of new regulations (even those aimed at the private sector) under 1995 legislation aimed at curbing "unfunded mandates." While this may not be the most patient or orderly process for learning how to do better cost-benefit analysis, it may be an effective way to induce regulatory agencies to take seriously their analytical responsibilities — assuming that Congress provides agencies with adequate staff and resources to undertake better analyses.

| How Should the Competing Risks of Regulation Be Considered?

A "sleeper issue" in the risk-management framework is how to identify, quantify, and weigh the risks (health and ecological) of regulatory action against the anticipated reductions in risks from the target pollution. The case studies reveal that these "competing risks" are currently addressed in a very uneven fashion.

In the case study of coke plants, the analysis commissioned by the EPA appears to have given insufficient consideration to the possibility that technology-based regulations, in conjunction with labor cost differentials and other factors, are gradually shifting coke production from the United States to China, Eastern Europe, and other developing countries. The global environmental implications of this trend cannot be good, particularly since the coke industry in China (now the largest coke producer in the world) is still using nineteenth-century technology (beehive ovens) in some cases.

The history of dry cleaning is characterized by a movement from one solvent to another and a discovery of competing risks from each new solvent. Perchloroethylene has some imperfections, but it looks awfully good when compared with most practical alternatives. Based on this historical experience, a new policy framework should induce regulators to be cautious about compelling use of new solvents in the dry cleaning industry unless the competing-risk issues are analyzed carefully. This lesson may be applicable to other industries as well.

The phaseout of lead in gasoline was not supported by a quantitative analysis of the alternative formulations likely to be investigated by refiners. It now appears that some increases in aromatics (such as benzene) may have occurred, thereby causing more smog and toxic emissions. It seems unlikely, however, that these competing risks are large enough to call into question the case for the accelerated phaseout of lead in gasoline. Thus, the lead case suggests that qualitative concerns about competing risks should not be used as an insurmountable hurdle for new regulations.

In the chlorofluorocarbon case study, policymakers took an aggressive stance on input substitution, but recognized that some new risks from substitutes might emerge. The EPA regulations call for careful toxicity testing of substitutes and complete reporting of results to the agency. However, the agency's analytic framework for weighing the competing risks of various substitutes is not fully worked out.

The clearest problem with competing risks was revealed by the case study of incineration technology. By all accounts, the current patchwork of environmental laws is causing incineration and landfilling to be regulated through independent procedures, with no coherent consideration of the relative risks of the two waste disposal methods. In Chapter 6 the authors make a plausible case that the country may be opening too many landfills and siting too few incinerators, from the perspective of risk minimization. The risk-management approach is rich enough to permit comparative-risk choices, so long as applicable laws and rules permit synoptic analyses. At the same time, this kind of approach would entail putting more thought into the scientific and value issues in comparing the risks of competing waste disposal technologies. This is a clear case of when the right answer cannot be found by common sense or intuition; technical analysis and reflection about value trade-offs are required.

| Prescription for Delay?

A serious concern of organized environmentalists is that the risk-management approach, by raising various evidentiary hurdles, will cause environmental policy to change too slowly in the face of real opportunities for environmental protection. On this issue, the case studies present some reason for concern.

The analysis supporting the 1985 lead phasedown was conducted in less than a year, but in this case there was no formal process of external review by a panel of independent experts from the scientific community. If this review had been required, the analytical phase might have taken up to two years. In the cases of perchloroethylene, dioxin, and coke oven emissions, the analytical processes (including multiple reviews by the EPA's Science Advisory Board) took almost a decade, as significant disagreements emerged between EPA staff and scientists on the board. The length of these delays lend credence to the concerns raised by David Doniger of the NRDC and others, that "risk-management" can be a prescription for endless debates among scientists.

At the same time, it must be remembered that scientists resist having their reputations abused on behalf of dubious regulatory proposals. There was strong scientific support for the EPA's actions on lead and chlorofluorocarbons, even though there remained significant scientific uncertainties. Peer-review requirements would probably not have slowed these decisions to anywhere near the point experienced in the perc, dioxin, and coke production cases. In each of the latter cases, there were serious scientific questions — questions that persist today — about whether the EPA's regulatory ambitions were based on significant public health or environmental problems.

The real issue is whether the precautionary principle should be invoked in such speculative cases. A possible solution in speculative cases might be to have the cost estimates for a regulatory proposal prepared and reviewed prior to the quantitative risk assessment. If regulatory costs are estimated to be small, a reviewing panel of scientists might be less reluctant to approve a risk-assessment report that makes numerous precautionary assumptions. When regulatory costs are likely to be large, a panel of scientific reviewers should probably insist on a more rigorous demonstration of risks to justify precautionary action.

This "cost first" process may have been occurring implicitly in the perc, dioxin, and coke cases, since the costs of the EPA's proposals in each of these situations were rather high, particularly compared with estimated benefits, and this reality may have been suspected by actors in the process. Even so, it is better to scrutinize cost claims in the open rather than allow them to influence the regulatory process via rumor (HGRR, 1995).

| Allocating Burdens of Proof

Given all of the above, the future of environmental policy seems to reside in a new paradigm that calls for more scientific and comparative analysis of the consequences of alternative technologies and regulatory alternatives. The analytic tools of risk assessment and cost-benefit analysis have an important role to play in the new paradigm, though analysts are far from being able to produce definitive answers with these tools.

A critical matter that has been raised by professional environmentalists concerns the proper allocation of burdens of proof under the new paradigm. The EPA shouldered the legal burden of proof in the decisions to accelerate the phaseout of lead in gasoline. It is interesting that the analysis supporting this decision survived considerable scientific scrutiny and was never challenged in court or overturned by federal judges. This case study suggests that it is possible to place analytical burdens on the EPA and still achieve strong regulatory decisions when the case is meritorious. In the case of chlorofluorocarbons, Congress gave the EPA the legal upper hand through statute, though the EPA's analysis would probably have been persuasive enough by itself to support strong regulatory action under a flexible cost-benefit standard.

Under the Clean Air Act Amendments of 1990, the EPA shouldered much smaller burdens of proof in the regulation of coke plants and dry cleaners. There is no requirement that the risks of toxic pollution be quantified or that the costs of technology-based controls be reasonable in light of benefits. The rules drafted by the EPA appear to be influenced primarily by what creative engineers believe can be accomplished at coke plants and dry cleaners without imposing undue financial burdens on the industry as a whole. Consumers will ultimately pay the price for these rules in higher prices for steel and dry cleaning, though people may never realize that the extra costs are attributable to EPA regulation. The problem with this approach is that it places the EPA in a poor position to defend what is being achieved (in health or environmental terms) for the considerable investments now being made in technological changes. If a regulation-weary Congress chooses to revisit the Clean Air Act Amendments of 1990, these rules will be very vulnerable to criticism.

When citizens are unwilling to devote tax dollars to agencies like the EPA to conduct careful risk analyses, it may make more sense to shift the analytic tasks to industry, with aggressive requirements for peer review of industry-sponsored analyses by independent experts. An approach worth considering would place the analytic burdens on industry as a precondition to obtaining permits to operate plants or permission to sell products that might pose risk. The FDA is under increasing pressure to make use of private, third-party reviews in pharmaceutical regulation, and Congress may consider using the same approach in environmental regulation. California's experience with Proposition 65, according to which companies are required to warn the public of chemical risks unless risk assessments suggest safety, is worthy of study as a possible model for future policy (Roe, 1989). If companies discover significant risks, then some form of environmental contracting with government that covers pollution to all affected media may be a more efficient approach to pollution reduction than simply tightening the current regulatory maze (Menell and Stewart, 1994).

Some environmentalists are concerned that more use of the risk-management approach will allow only incremental changes to be considered in industrial processes, rather than fundamental ones. While there may be political obstacles to making fundamental changes to current processes, it certainly seems analytically possible to compare diverse technologies such as landfilling and incineration. Indeed, the EPA has already published preliminary analyses comparing pollutant emissions from electric cars to those resulting from continued use of the internal combustion engine. And it should be noted that some of the earliest applications of comparative risk analysis entailed comparisons of the mortality impacts of alternative energy sources (such as coal versus nuclear power). There is nothing about the analytical tools that prevents investigation of fundamentally different technologies, and therefore legislators should work hard to legitimize these options within the risk-management framework.

Considering more radical technological alternatives does offer a greater challenge to analysts. The costs of big changes may be more uncertain than the costs of incremental ones. More diverse types of environmental impacts (for example, a slower rate of global warming as well as smog prevention from a tax on fossil fuels) may need to be considered. But risk assessment and to a lesser extent cost-benefit analysis, as shown by the cases in this book, can support aggressive regulation and process change and are not necessarily in conflict with the aspirations of ambitious environmentalists. If a strong case can be made for more radical alternatives based on pragmatic impacts on human health and ecosystems, then the available analytical tools — while requiring further development — seem to be flexible enough to meet the challenge.

| Sources

Cross, F. B., 1990. *Legal Responses to Indoor Air Pollution* (New York: Quorom Books).

———, 1994. "The Public Role in Risk Control," *Environmental Law* 24:888–969.

Doniger, D., 1989. National Clean Air Coalition, "Clean Air Act Amendments of 1989," *Hearings before the Senate Committee on Environment and Public Works,* 101st Cong., 1st sess., September 21, 28–30.

Environmental Protection Agency (EPA), 1984. *Risk Assessment and Management: Framework for Decision Making* (Washington, D.C.: Environmental Protection Agency, December).

———, 1987. *Unfinished Business: A Comparative Assessment of Environmental Protection* (Washington, D.C.: Environmental Protection Agency).

———, 1990a. *Reducing Risk: Setting Priorities and Strategies for Environmental Protection,* Science Advisory Board (Washington, D.C.: Environmental Protection Agency).

———, 1990b. *Environmental Investments: The Cost of a Clean Environment,* a Summary, December (Washington, D.C.: Environmental Protection Agency).

Finkel, A. M., and D. Golding (eds.), 1995. *Worst Things First? The Debate Over Risk-Based National Environmental Priorities* (Baltimore: Johns Hopkins University Press).

Freeman, A. M., 1993. *The Measurement of Environmental and Resource Values: Theory and Methods* (Washington, D.C.: Resources for the Future).

Graham, J. D., 1995. "Comparing Opportunities to Reduce Health Risks: Toxin Control, Medicine and Injury Prevention" (Washington, D.C.: National Center for Policy Analysis).

Graham, J. D., and D. R. Holtgrave, 1990. "Coke Oven Emissions: A Case Study of Technology-Based Regulation," *Risk: Issues in Health and Safety* 1:243–272.

Hahn, R. W., and J. A. Hird, 1991. "The Costs and Benefits of Regulation: Review and Synthesis," *Yale Journal on Regulation* 8:233–279.

Harvard Group on Risk Management Reform (HGRR), 1995. "Special Report: Reform of Risk Regulation: Achieving More Protection at Less Cost," *Human and Ecological Risk Assessment* 1(3):183–206.

Holland, C. D., and R. L. Sielken, 1993. *Quantitative Cancer Modelling and Risk Assessment* (Englewood Cliffs, N.J.: Prentiss-Hall).

Kelman, S., 1981. "Cost-Benefit Analysis: An Ethical Critique," *Regulation* January/February.

Landy, M. K., M. J. Roberts, and S. R. Thomas, 1990. *EPA: Asking the Wrong Questions* (New York: Oxford University Press).

Leonard, H. B., and R. J. Zeckhauser, 1986. "Cost-Benefit Analysis Applied to Risks: Its Philosophy and Legitimacy," in D. MacLean, ed., *Values at Risk.*

McGarity, T. O., 1991. *Reinventing Rationality: The Role of Regulatory Analysis in the Federal Bureaucracy* (Cambridge: Cambridge University Press).

Menell, P. S., and R. B. Stewart, 1994. *Environmental Law and Policy* (Boston: Little, Brown) pp. 420–421.

Moynihan, D. P., 1993. "Environmental Risk Reduction Act," *Congressional Record,* January 21.

National Academy of Public Administration (NAPA), 1995. *Setting Priorities, Getting Results,* Summary Report, April (Washington, D.C.: National Academy of Public Administration).

National Research Council (NRC), 1994. *Science and Judgement in Risk Assessment* (Washington, D.C.: National Academy Press).

Reilly, W., 1994. Address to the John F. Kennedy School of Government, Harvard University, December 13.

Roe, D., 1989. "An Incentive-Conscious Approach to Toxic Chemical Controls," *Economic Development Quarterly* 3(3):179–187.

Ruckelshaus, W., 1983. "Science, Risk, and Public Policy," *Science* 221:1026–1028.

———, 1985. "Risk, Science, and Democracy," *Issues in Science and Technology* 1(3):19–38.

Tolley, G., D. Kenekel, and R. Fabian, 1994. *Valuing Health for Policy: An Economic Approach* (Chicago: University of Chicago Press).

INDEX

Page numbers in *italic* refer to figures and tables.